The Mathematics of Harmony ... Hilbert's Fourth Problem

Alexey Stakhov
Samuil Aranson

The Mathematics of Harmony and Hilbert's Fourth Problem

The Way to the Harmonic Hyperbolic and Spherical Worlds of Nature

LAP LAMBERT Academic Publishing

Impressum / Imprint

Bibliografische Information der Deutschen Nationalbibliothek: Die Deutsche Nationalbibliothek verzeichnet diese Publikation in der Deutschen Nationalbibliografie; detaillierte bibliografische Daten sind im Internet über http://dnb.d-nb.de abrufbar.

Alle in diesem Buch genannten Marken und Produktnamen unterliegen warenzeichen-, marken- oder patentrechtlichem Schutz bzw. sind Warenzeichen oder eingetragene Warenzeichen der jeweiligen Inhaber. Die Wiedergabe von Marken, Produktnamen, Gebrauchsnamen, Handelsnamen, Warenbezeichnungen u.s.w. in diesem Werk berechtigt auch ohne besondere Kennzeichnung nicht zu der Annahme, dass solche Namen im Sinne der Warenzeichen- und Markenschutzgesetzgebung als frei zu betrachten wären und daher von jedermann benutzt werden dürften.

Bibliographic information published by the Deutsche Nationalbibliothek: The Deutsche Nationalbibliothek lists this publication in the Deutsche Nationalbibliografie; detailed bibliographic data are available in the Internet at http://dnb.d-nb.de.

Any brand names and product names mentioned in this book are subject to trademark, brand or patent protection and are trademarks or registered trademarks of their respective holders. The use of brand names, product names, common names, trade names, product descriptions etc. even without a particular marking in this works is in no way to be construed to mean that such names may be regarded as unrestricted in respect of trademark and brand protection legislation and could thus be used by anyone.

Coverbild / Cover image: www.ingimage.com

Verlag / Publisher:
LAP LAMBERT Academic Publishing
ist ein Imprint der / is a trademark of
OmniScriptum GmbH & Co. KG
Heinrich-Böcking-Str. 6-8, 66121 Saarbrücken, Deutschland / Germany
Email: info@lap-publishing.com

Herstellung: siehe letzte Seite /
Printed at: see last page
ISBN: 978-3-659-52803-3

Preface

Non-Euclidean geometry. In the first half of the 20th century, the outstanding mathematical discovery in geometry was made. Russian geometer **Nikolay Lobachevski** (1792-1856) introduced the so-called "imaginary geometry," which changed our ideas about the geometry of the physical world. This geometry was called **"hyperbolic geometry"** [1] on the grounds of the fact that the classical hyperbolic functions were used to describe the mathematical relations of this geometry.

The following remarkable words belong to the creator of the hyperbolic geometry Nikolay Lobachevski:

"Mathematicians drew all their attention on the Advanced Parts of Analysis, by neglecting the origin of Mathematics and not wishing to dig in the field, which they already went through and had left behind."

Unlike many mathematicians, Lobachevski did not hesitate to "dig" into some of the unsolved problems of ancient mathematics, in particular, into the problem of Euclid's 5th postulate and this study led him to the outstanding mathematical discovery - the creation of hyperbolic geometry [1].

Lobachevski begins his famous *"Geometric study on the theory of parallel lines"* with the following words:

"In geometry, I found some imperfections, which, in my opinion, are the reason why this science until now is in the state, in which it had come to us from Euclid. I attribute to these imperfections the following: the vagueness in the first definitions of geometric quantities, the methods of measurement of these quantities, and, finally, the important gap in the theory of parallel lines."

The history of hyperbolic geometry was accompanied with many dramatic events. The first of them is a very negative reaction of the Russian academic science on the discovery of "Kazanian rector" Nikolay Lobachevski. Hounding began with the negative review on Lobachevski's work, written by the famous Russian mathematician academician Ostrogradski. Only thanks to the support of the outstanding German mathematician Gauss, the hyperbolic geometry has got the deserved recognition among 19 century's mathematicians. Academician Kolmogorov in his excellent book [2] pointed out that *"Lobachevski's geometry was the most important discovery of the early of 19th century."*

Independently from Lobachevski, Hungarian mathematician Janos Bolyai (1802-1860) came to the same ideas, however, he published 3 years later his researches on hyperbolic geometry. Later it was found from Gauss' correspondence that the outstanding German mathematician **Carl Friedrich Gauss** (1777-1855) came to the same revolutionary ideas in hyperbolic geometry independently from Lobachevski and Bolyai.

The works by Lobachevski, Bolyai and Gauss stimulated the creation of new classes of "non-Euclidean geometries." According to [3], the non-Euclidean geometry is any geometric system that is different from Euclidean geometry, but traditionally, the term "non-Euclidean geometry" is used more narrowly and relates to the two geometric systems: **hyperbolic geometry** [1] and **spherical geometry** [4,5]. As Euclidean geometry, these geometries relate to the metric geometries of the space with constant curvature. Zero curvature corresponds to Euclidean geometry, positive curvature to the spherical geometry, negative curvature to the hyperbolic geometry.

A form of the metric for homogeneous planimetry depends on the selected system of curvilinear coordinates, below we consider the formulas for the metrics for the case of semi-geodesic coordinates:

- Euclidean geometry: $(ds)^2 = (dx)^2 + (dy)^2$ (Pythagoras Theorem).

- Spherical geometry: $(ds)^2 = (dx)^2 + \cos^2\left(\dfrac{x}{R}\right)(dy)^2$. Here, R is a radius of sphere.

- Lobachevski's geometry: $(ds)^2 = (dx)^2 + ch^2\left(\dfrac{x}{R}\right)(dy)^2$. Here, R is a radius of the curvature of Lobachevski's plane, ch is a hyperbolic cosine.

Hilbert's Problems. According to Kolmogorov [2], Lobachevski's hyperbolic geometry became the starting point of the modern stage in the development of mathematics. To stimulate the development of this mathematical discipline in the 20th century, David Hilbert drew a special attention to the hyperbolic geometry in his 23 mathematical problems [6,7]. He had formulated a very important problem (Hilbert's Fourth Problem) for finding of new hyperbolic geometries, which are close to Euclidean geometry. Unfortunately, this problem is still not resolved. Currently, most mathematicians are inclined to believe that Hilbert's Fourth Problem has been formulated too vague what makes complicated its final solution [8]. That is, the mathematicians of the 20 century laid the blame for the failure in the solution of this problem on Hilbert himself. The purpose of this book is a presentation of new and original solution of Hilbert's Fourth Problem.

In this book we develop a new approach to the solution of Hilbert's Fourth Problem, called "the game of functions." This approach has only been possible in the 21st century, when a general theory of "harmonic" hyperbolic functions, based on the ancient "golden ratio" and its generalization – the "metallic proportions," has been developed [9-14]. The uniqueness of these functions is the fact that they are inseparably connected with the Fibonacci numbers and their generalization - Fibonacci λ-numbers (λ>0 is a given real number). Each of these new classes of hyperbolic functions, the number of which is theoretically infinite, "generates" a new hyperbolic geometry that has new geometric and recursive properties. In this book we spread this approach on **spherical geometry**. With this purpose we have introduced the so-called **spherical Fibonacci functions**, which underlie **spherical solution of Hilbert's Fourth Problem**.

The Mathematics of Harmony. Euclid's *Elements* is a source for the development of another important mathematical discipline - "the Mathematics of Harmony." As follows from Proclus' hypotheses [15-18], Euclid wrote his famous *Elements* with the purpose to set forth a complete theory of the Platonic solids, which had symbolized in ancient Greece the Universe Harmony. To create this theory, Euclid introduced the task of dividing the line segment into extreme and mean ratio (the "golden ratio") in the Book II. Proclus' hypothesis underlies Alexey Stakhov's book "The Mathematics of Harmony. From Euclid to Contemporary Mathematics and Computer Science" (World Scientific, 2009) [19]. The publication of this book is a reflection of one of the most important trends in the development of modern science. The essence of this trend is to return back to the "harmonic ideas" of Pythagoras and Plato (the "golden ratio" and Platonic solids), embodied in Euclid's "Elements" [18].

The "golden" and "silver" hyperbolic geometries. A new geometric theory of phyllotaxis has been recently developed by Ukrainian architect Oleg Bodnar [20-23]. This geometry is based on the "golden" hyperbolic functions with the base $\Phi = \left(1+\sqrt{5}\right)/2 \approx 1.618$ (the "golden ratio) and called the "golden" hyperbolic geometry. To assess the proximity of new hyperbolic geometries to Lobachevski's classical geometry with the base $e \approx 2.71$, the important concept of the "distance" has been introduced in the book. It is proved that the "silver" hyperbolic functions with the base $1+\sqrt{2} \approx 2.41$ "generates" new ("silver") hyperbolic geometry, which has minimal distance to Lobachevski's classical

5

geometry. This fact allows us to put forward the hypothesis that the "silver" hyperbolic geometry may be widespread in real physical world.

The newest scientific discoveries. The newest discoveries in chemistry and crystallography: **fullerenes** [24], based on the "truncated icosahedron" (Nobel Prize - 1996), and **quasi-crystals** [25], based on the icosahedral or pentagonal symmetry (Nobel Prize - 2011), are brilliant confirmation of the revival of the ancient "harmonic ideas" (Pythagoras, Plato, Euclid) in modern science.

A number of such discoveries in modern science are increasing continuously. These include: **"the law of structural harmony of systems"** by Edward Soroko [26], based on the generalized golden proportions, the **"law of spiral biosimmetry transformation"** by Oleg Bodnar [20-23], based on the "golden" hyperbolic functions. It also includes the original solution of **Hilbert's Tenth Problem** [27], based on the Fibonacci numbers (the author is the Russian mathematician Yuri Matiyasevich), and also a new theory of the genetic code, based on the **"golden genomatrices"** (the author is the Russian researcher Sergey Petoukhov, Moscow) [28]. These examples could go on.

Thus, the modern mathematics and theoretical natural sciences begun to use widely the "harmonic ideas" by Pythagoras, Plato and Euclid. And we have every right to talk about the "revival" of ancient Greeks' "harmonic ideas" in modern science. This fact puts forward the problem of the revival of these ancient harmonic ideas in modern mathematics. The book [19] is the answer of contemporary mathematics for this important trend.

Relationships between Mathematics of Harmony and Hyperbolic Geometry. As is showed in [18], the mathematics of harmony is dating back to Euclid's *Elements* in its origin. This means that Euclid's *Elements* became the source of two mathematical directions – the **"Non-Euclidean Geometry,"** emerged in the 19th century, and the **"Mathematics of Harmony,"** emerged in the 21st century. As is showed in [18], there are very interesting mathematical relationships between them. They can significantly affect on the future development of non-Euclidean geometry. We are talking about the following scientific results, obtained in the "Mathematics of Harmony" [19]:

1. Fibonacci hyperbolic trigonometry, based on the "golden ratio" and Fibonacci numbers [9-12].
2. A new geometric theory of phyllotaxis ("Bodnar's geometry") [20-23].

3. The general theory of hyperbolic functions, based on the "metallic proportions" [13, 14].

4. The original solution of Hilbert's Fourth Problem for the hyperbolic geometry [29-32].

The goals of the book. The present book is based, first of all, on the works [9-14, 18-20, 29-32] and pursues different goals. The first goal is to develop a new view on Euclid's *Elements* based on Proclus' hypothesis [15-18] what changes our ideas about Euclid's *Elements* and mathematics history, starting from Euclid. The next goal is to give an overview of the main mathematical results of the "mathematics of harmony" [19] as a new interdisciplinary direction of modern science. Another goal is to set forth a theory of hyperbolic and spherical Fibonacci functions, which were obtained in the framework of the mathematics of harmony [19]. However, the main goal is to set forth the original solution of Hilbert's Fourth Problem for hyperbolic and spherical geometries what opens new ways in the development of non-Euclidean geometry and all theoretical natural sciences.

Chapter 1

THE MATHEMATICS OF HARMONY, PROCLUS HYPOTHESIS, AND THE GOLDEN RATIO

1.1. What is the Mathematics of Harmony?

1.1.1. The opinion of Academician Yury Mitropolski

What is the Mathematics of Harmony"? What is its role in modern science and mathematics? The outstanding Ukrainian mathematician, the leader of the Ukrainian School of Mathematics, Honorary Director of the Institute of Mathematics of the Ukrainian Academy of Sciences and Editor-in Chief of the "Ukrainian Mathematical Journal" Academician Yury Mitropolski wrote the following in his commentary [33]:

"One may wonder what place in the general theory of mathematics is occupied by Mathematics of Harmony created by Prof. Stakhov? It seems to me, that in the last centuries, as Nikolay Lobachevski said, "mathematicians drew all their attention on the Advanced Parts of Analysis, neglecting the origins of Mathematics and not wishing to dig in the field, which they already passed and had left behind." As a result, this created a gap between "Elementary Mathematics" – the basis of modern mathematical education, and "Advanced Mathematics". In my opinion, the Mathematics of Harmony developed by Prof. Stakhov fills up that gap. That is, the "Mathematics of Harmony" is a big theoretical contribution, first of all to the development of "Elementary Mathematics" and therefore it should be considered as having great importance for mathematical education."

Figure 1.1. Academician Yury Mitropolski (1917 – 2008)

Thus, according to Mitropolski, the Mathematics of Harmony is a new mathematical discipline, which fills the gap between the "Elementary Mathematics" and "Higher Mathematics." That is, this new theory puts forward new tasks in the field of the "Elementary Mathematics." In its origins, this theory goes back to the ancient mathematical topics: *measurement theory, number theory, numeral systems, elementary functions* " and so on.

The main challenge of the "Mathematics of Harmony" is to find new results in the field of the "Elementary Mathematics," based on the "golden ratio" and Fibonacci numbers. It is proved a high efficiency of these results in such areas as theory of recurrence relations, theory of elementary functions, hyperbolic geometry, and finally, the computing and measuring engineering, coding theory.

1.1.2. About the term of "the Mathematics of Harmony"

For the first time, the term of "the Mathematics of Harmony" was used in the article "Harmony of spheres," placed in «The Oxford dictionary of philosophy» [34]:

*"**Harmony of spheres.** A doctrine often traced to Pythagoras and fusing together mathematics, music, and astronomy. In essence the heavenly bodies being large objects in motion, must produce music. The perfection of the celestial world requires that this music be harmonious, it is hidden from our ears only because it is always present. **The mathematics of harmony was a central discovery of immense significance to the Pythagoreans.**"*

Thus, the concept of "the Mathematics of Harmony" is associated here with the "harmony of the spheres," which is also called the "harmony of the world» (Latin "harmonica mundi") or world music (Latin "musical mundane"). The harmony of the spheres [35] is the ancient and medieval doctrine about the musical-mathematical organization of Cosmos that goes back to the Pythagorean and Platonic philosophical tradition.

Another mention about "the Mathematics of Harmony," as the ancient Greek great discovery, we find in the book by Vladimir Dimitrov. "A new kind of social science. Study of self-organization of human dynamics" [36]. Let us consider the quote from the book:

*"Harmony was a key concept of the Greeks, a conjunction of three strands of meaning. Its root meaning was **aro**, join, so **harmonia was what joined**. Another meaning was proportion, the balance of things that allowed an easy fit. The quality of joining and proportion then came to be seen in music and other arts.*

*The precondition for harmony for the Greeks was expressed in the phrase "nothing to much". It also had a mysterious positive quality, which became the object of enquiry of their finest minds. Thinkers such as Pythagoras sought to capture the mystery of harmony as something both inexpressible yet also illuminated by mathematics. **The mathematics of harmony** explored by the ancient Greeks is still an inspiring model for contemporary scientists. Crucial to it is their discovery of its quantitative expression in astonishing diversity and complexity of nature through the golden mean (golden ratio), Φ (phi): $\Phi = \dfrac{1+\sqrt{5}}{2}$, which is approximately equal to 1.618. It is described by Euclid in book five of his Elements: "A straight line is said to have been cut in extreme and mean ratio when, as the whole line is to greater, so is greater to the less".*

Thus, in the book [36] the concept of the mathematics of harmony is directly associated with the *golden ratio* – the most important ancient mathematical discovery in the field of harmony, which at that time was called *the division of the segment in extreme and mean ratio.*

Finally, it is pertinent to mention that this term was used in Stakhov's speech «The Golden Section and Modern Harmony Mathematics», made at the 7th International Conference «Fibonacci Numbers and Their Applications" (Austria, Graz, 1996) [37].

1.2. The most important periods in the development of the Mathematics of Harmony

As is mentioned above, very important definition of the "Mathematics of Harmony" was introduced in [34, 36] to emphasize the most important feature of the ancient Greek's science (a study of the Universe Harmony). The greatest interest in the "Universe Harmony," that is, in the ideas of Pythagoras, Plato and Euclid, always arose in the periods of greatest prosperity of the "human spirit." From this point of view, in the studying of the "Mathematics of Harmony" we can point the following critical periods.

1.2.1. The ancient Greek period

Conventionally, it can be assumed that this period starts with Pythagoras and Plato. Euclid' *Elements* became a final event of this important period. According to Proclus' hypothesis [15-18], Euclid created his *Elements* in order to create the complete geometric theory of the five "Platonic solids," which have been associated in the ancient Greek science with the "Universe Harmony." The geometric fundamentals of the theory of Platonic solids (Fig.1.2) were described by Euclid in the concluding Book (Book XIII) of the *"Elements."*

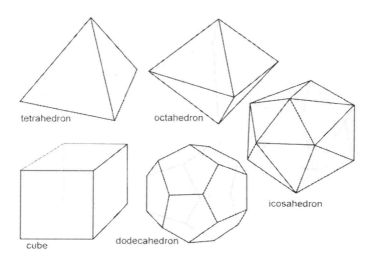

Figure 1.2. Platonic solids: tetrahedron, octahedron, cube, dodecahedron, icosahedron

In addition, Euclid simultaneously had introduced here some advanced achievements of ancient Greek mathematics, in particular, the "golden ratio" (Book II), which was used by Euclid for the creation of the geometric theory of the *Platonic solids,* in particular, *dodecahedron.*

1.2.2. The Middle Ages

In the Middle Ages, very important mathematical discovery was made in the field. The famous Italian mathematician **Leonardo of Pisa (Fibonacci)** wrote the book "Liber Abaci" (1202). In this book, he described *"the task of rabbits reproduction."* By solving this problem, he found the remarkable numerical sequence – the *Fibonacci numbers* F_n:

$$1,1,2,3,5,8,13,21,34,55,89,..., \qquad (1.1)$$

which are given by the recurrence relation:

$$F_n = F_{n-1} + F_{n-2}; \quad F_1 = F_2 = 1 \qquad (1.2)$$

13

1.2.3. The Renaissance

This period is connected with the names of the prominent figures of the Renaissance: **Piero della Francesca** (1412-1492), **Leon Battista Alberti** (1404-1472), **Leonardo da Vinci** (1452-1519), **Luca Pacioli** (1445-1517), **Johannes Kepler** (1571-1630). In that period two books, which were the best reflection of the idea of the "Universe Harmony," were published. The first of them is the book "Divina Proprtione" («The Divine Proportion") (1509). This book was written by the outstanding Italian mathematician and scholar monk **Luca Pacioli** under the direct influence of **Leonardo da Vinci**, who illustrated Pacioli's book.

Also the brilliant astronomer of 17th century **Johannes Kepler** made an enormous contribution to the development of the "harmonic ideas" of Pythagoras, Plato and Euclid.

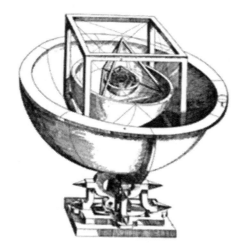

Figure 1.3. Kepler's Cosmic Cup

In his first book *Mysterium Cosmographicum* (1596) he built the so-called *"Cosmic Cup"* (Fig.1.3), the original model of the Solar system, based on the Platonic solids. The book *Harmonice Mundi* (Harmony of the Worlds) (1619) is the main Kepler's contribution into the Doctrine of the Universe Harmony. In the *Harmony*, he attempted to explain the proportions of the Universe — particularly the astronomical and astrological aspects — by using musical terms. The

"Musica Universalis" or "Music of the Spheres," which had been studied by Pythagoras and Ptolemy, was the central idea of Kepler's *Harmony*.

1.2.4. The 19th century

This period is connected with the works of the French mathematicians **Jacques Philip Marie Binet** (1786-1856), **Francois Edouard Anatole Lucas** (1842-1891), German poet and philosopher **Adolf Zeising** (born in 1810) and the German mathematician **Felix Klein** (1849 – 1925).

Jacques Philip Marie Binet derived a mathematical formula to represent the Fibonacci numbers through the "golden ratio" $\Phi = \dfrac{1+\sqrt{5}}{2}$.

Francois Edouard Anatole Lucas introduced the Lucas numbers L_n, which are calculated by the some recurrence relation as the Fibonacci numbers (1.1), but with other seeds:

$$L_n = L_{n-1} + L_{n-2}; \quad L_1 = 1, L_2 = 3 \qquad (1.3)$$

The recurrence relation (1.3) generates the following numerical sequence:

$$L_n : 1, 3, 4, 7, 11, 18, 29, \ldots \qquad (1.4)$$

The merit of Binet and Lucas is the fact that their researches became a launching pad for Fibonacci researches in the Soviet Union, the United States, Britain and other countries [38-40].

German poet Adolf Zeising in 1854 published the book *«Neue Lehre von den Proportionen des menschlichen Körpers aus einem bisher unerkannt gebliebenen, die ganze Natur und Kunst durchdringenden morphologischen Grundgesetze entwickelt»*. The basic Zeising's idea is to formulate the Law of proportionality. He formulated this Law as follows:

"A division of the whole into unequal parts is proportional, when the ratio between the parts is the same as the ratio of the bigger part to the whole, this ratio is equal to the golden mean".

The famous German mathematician Felix Klein in 1884 published the book *"Lectures on the icosahedron"* [41], dedicated to the geometric theory of the icosahedron and its role in the general theory of mathematics. Klein treats the icosahedron as a mathematical object, which is a source for the five mathematical theories: *geometry, Galois theory, group theory, invariant theory and differential equations.*

What is a significance of Klein's ideas from the standpoint of the Mathematics of Harmony [19]? According to Klein, the Platonic icosahedron, based on the "golden ratio," is the main geometric figure of mathematics. It follows from this that the "golden ratio" is the main geometric object of mathematics, which, according to Klein, can unite all mathematics.

This Klein's idea is consistent with the ideas of the article *"Generalized Golden Sections and a new approach to geometric definition of a number,"* published by Alexey Stakhov in the "Ukrainian Mathematical Journal" [42] according to the recommendation of academician Mitropolsky. This article presents the concept of the "golden" number theory, which can be the basis for the "golden" mathematics, based on the "golden ratio" and its generalizations.

1.2.5. The first half of the 20th century

In the first half of the 20th century, the development of the "golden" paradigm of the ancient Greeks is associated with the names of the Russian Professor of architecture **G.D.Grimm** (1865-1942) and the classic of the Russian religious philosophy **Paul Florensky** (1882-1937).

In the theory of architecture, it is well-known the book "Proportionality in Architecture," published by Prof. Grimm in 1935 [43]. The purpose of the book has been formulated in the "Introduction" as follows:

"In view of the exceptional significance of the Golden Section in the sense of the proportional division , which establishes a permanent connection between the whole and its parts and gives a constant ratio between them (which is unreachable by any other division), the scheme, based on it, is the main standard and is accepted by us in the future as a basis for checking the proportionality of historical monuments and modern buildings ... Taking this general importance of the Golden Section in all aspects of architectural thought , the theory of proportionality, based on the proportional division of the whole into parts corresponding to the Golden Section, should be recognized as the architectural basis of proportionality at all."

In the 20th years of 20th century, Paul Florensky wrote the work *"At the watershed of a thought."* Its third chapter is devoted to the "golden ratio". The

Belorussian philosopher Edward Soroko in the book [26] evaluates Florensky's work as follows:

" *The aim was to derive analytically the stability of the whole object , which is in the field of the effect of oppositely oriented forces. The project was conceived as an attempt to use the "golden ratio and its substantial basis, which manifests itself not only in a series of experimental observations of nature, but on the deeper levels of knowledge, for the case of penetration into the dialectic of movement, into the substance of things.* "

1.2.6. The second half of the 20th century and the 21st century

In the second half of 20th century the interest in this area is increasing in all areas of science, including mathematics. The Soviet mathematician **Nikolay Vorobyov** (1925-1995) [38], the American mathematician **Verner Hoggatt** (1921-1981) [39], the English mathematician **Stefan Vajda** [40] and others became the most outstanding representatives of this direction in mathematics.

The revival of the idea of harmony in modern science is determined by new scientific realities. The penetration of the Platonic solids, the "golden ratio" and Fibonacci numbers in all areas of theoretical natural sciences (crystallography, chemistry, astronomy, earth science, quantum physics, botany, biology, geology, medicine, genetics, etc.), as well as in computer science and economics was the main reason for the renewed interest in the ancient idea of the Universe Harmony in modern science and the stimulus for the development of the "Mathematics of Harmony."

1.3. Proclus hypothesis: a new view on Euclid's "Elements" and the history of mathematics

1.3.1. Three "key" problems of the ancient mathematics

"Proclus hypothesis," formulated in the 5th century AD by the famous Greek philosopher and mathematician **Proclus Diadochus** (412 – 485), contains the unexpected view on Euclid's *Elements*.

According to Proclus, Euclid's goal was not to set forth the geometry itself, but to build the complete theory of regular polyhedra ("Platonic solids"). This theory was outlined by Euclid in the XIII-th, that is, the concluding book of the *Elements* what in itself is an indirect confirmation of the Proclus hypothesis.

To solve this problem, Euclid had included the necessary mathematical information into the *Elements*. The most curious thing is that he had introduced in the Book II the "golden section," used by him for the creation of geometric theory of the dodecahedron. In Plato's Cosmology, the regular polyhedra had been associated with the "Harmony of the Universe." This means, that Euclid's *Elements* are based on the "harmonic ideas" of Pythagoras and Plato, that is, Euclid's *Elements* are historically first variant of the "Mathematics of Harmony." This unexpected view on the *"Elements"* leads to the conclusion, which changes our view on the history and structure of mathematics.

As is known, the famous Russian mathematician academician Kolmogorov in the book [2] had identified the two main, that is, "key" problems, which stimulated the development of mathematics at the stage of its origins: the *counting problem* and the *measurement problem*. However, it follows from "Proclus hypothesis" another "key" problem: the *harmony problem*, which underlies Euclid's *Elements*.

We can see that three "key" problems – the *counting problem*, the *measurement problem*, and the *harmony problem* – underlie the origins of mathematics (see Fig.1.4).

The first two "key" problems resulted in the creation of two fundamental notions of mathematics – *natural number* and *irrational number*, which underlie the **Classical Mathematics**. The harmony problem, connected with the division in extreme and mean ratio (the "golden ratio") (Proposal II.11 of Euclid's *Elements*) resulted in the **Harmony Mathematics** [19] – a new interdisciplinary direction of contemporary science and mathematics.

This approach leads to the unexpected conclusion for many mathematicians. It turns out, that in parallel with the **Classical Mathematics**, another mathematical direction – the **Mathematics of Harmony** – was developing in mathematics, starting since ancient Greek period.

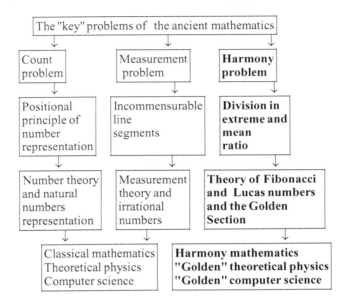

Figure 1.4. Three "key" problems of the ancient mathematics

Similarly to the Classical Mathematics, the Mathematics of Harmony takes its origin in Euclid's *Elements*. However, the Classical Mathematics focuses on the axiomatic approach and other ancient achievements (number theory, theory of irrationalities and so on), while the Mathematics of Harmony is based on the "golden ratio" (Proposal II.11) and Platonic Solids described in Book XIII of Euclid's *Elements*. Thus, Euclid's *Elements* is a source of two independent mathematical directions: the **Classical Mathematics** and the **Mathematics of Harmony**.

For many centuries, the creation of the Classical Mathematics, the Queen of Natural Sciences, was the main focus of mathematicians. However, starting from Pythagoras, Plato, Euclid, Pacioli, Kepler, the intellectual forces of many prominent mathematicians and thinkers were directed towards the development of the basic concepts and applications of the Mathematics of Harmony. We have no right to negate this important fact in the history of mathematics.

Unfortunately, these two important mathematical directions (Classical Mathematics and Mathematics of Harmony) evolved separately from one other. A time came to unite these important mathematical directions. This unusual union can lead to new scientific discoveries in mathematics and theoretical natural sciences.

A new approach to the origin of mathematics (see Fig.1.3) is very important for mathematical education. This approach introduces, in very natural manner, the idea of harmony and the golden ratio into mathematical education. This provides for pupils and students the access to ancient science and to its main achievement – the harmonic ideas – and to tell them about the most important architectural and sculptural works of the ancient art based upon the golden ratio (including pyramid of Khufu (Cheops), Nefertiti, Parthenon, Doryphoros, Venus and so on).

1.3.2. A discussion of Proclus' hypothesis in the historical-mathematical literature

The analysis of Proclus' hypothesis is found in many mathematical sources. Consider some of them [15-17]. In the book [15] we read: *"According to Proclus, the main objective of the "Elements" was to present the geometric construction of the so-called Platonic solids."*

In the book [16], this idea got a further development: *"Proclus, by mentioning all previous mathematicians of Plato's circle, said: "Euclid lived later than the mathematicians of Plato's circle, but earlier than Eratosthenes and Archimedes ... He belonged to Plato's school and was well acquainted with Plato's philosophy and his cosmology; that's why he put a creation of the geometric theory of the so-called Platonic solids as the main purpose of the Elements."*

This comment is very important and draws our attention to the connection of Euclid with Plato. Euclid fully shared Plato's philosophy and cosmology, based on Platonic solids, that is why, Euclid put forward the creation of the geometric theory of Platonic solids as the main purpose of the *Elements*.

In the book [17], there is discussed the influence of Plato and Euclid' ideas on Johannes Kepler at designing of Kepler's *Cosmic Cup* (Fig.1.3) in his first book "Mysterium Cosmographicum":

"Kepler's project in "Mysterium Cosmographicum" was to give "true and perfect reasons for the numbers, quantities, and periodic motions of celestial orbits." The perfect reasons must be based on the simple mathematical principles, which had been found by Kepler in the Solar system, by using multiple geometric demonstrations. The general scheme of his model was borrowed by Kepler from Plato's Timaeus, but the mathematical relations for the Platonic solids (pyramid, cube, octahedron, dodecahedron, icosahedron) were taken by Kepler from the works by Euclid and Ptolemy. At that, Kepler followed to Proclus and believed that "the main goal of Euclid was to build a geometric theory of the so-called Platonic solids." Kepler was completely fascinated by Proclus, he often quotes him and calls him "Pythagorean."

From this quote, we can conclude that Kepler used the Platonic solids to create the *Cosmic Cup*, but all the mathematical relations for the Platonic solids were borrowed by him from Book XIII of the *Elements,* that is, he united in his studies Plato's Cosmology with Euclid's *Elements*. At that, he fully believed in Proclus' hypothesis that the main goal of Euclid was to give the complete geometric theory of Platonic solids, which were used by Kepler in his geometric model of the Solar system (Fig.1.3).

1.4. The Golden Ratio in Euclid's Elements

1.4.1. Proposition II.11

In Euclid's *"Elements"* we meet the task, which later had played a great role in the development of science. This task is called the *"division of line segment in extreme and mean ratio."* In the *"Elements"* this task occurs in two forms. The first form is formulated in the Proposition 11 of the Book II.

Proposition II.11. Divide a given line segment *AD* into two unequal parts *AF* and *FD* so that the area of the square, which is built on the larger segment *AF* would be equal to the area of the rectangle, which is built on the segment *AD* and the smaller segment *FD*.

Depict this problem geometrically (Fig.1.5).

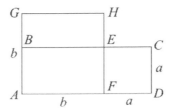

Figure 1.5. A division of a line segment in extreme and mean ratio (the "golden ratio")

Thus, according to the Proposition II.11, the area of the square *AGHF* should be equal to the area of the rectangle *ABCD*. If we denote the length of the larger segment *AF* through *b* (it is equal to the side of the square *AGHE*), and the side of the smaller segment *FD* through *a* (it is equal to the vertical side of the rectangle *ABCD*), then the condition for the Proposition II.11 can be written as follows:

$$b^2 = a \times (a+b).\qquad(1.5)$$

In Euclid's *Elements,* we meet another form of the "golden ratio." This form follows from the first one, given by (1.5), if we make the following transformations. Dividing both parts of (1.5) at first by *a*, and then by *b*, we get the following proportion:

$$\frac{b}{a} = \frac{a+b}{b}.\qquad(1.6)$$

The proportion (1.6) has the following geometric interpretation (Fig.1.5). Divide the segment *AB* in the point *C* in such a relation when the larger part *CB* relates to the smaller part *AC*, as the segment *AB* to its larger part *CB* (Fig.1.6), that is,

$$\frac{AB}{CB} = \frac{CB}{AC}.\qquad(1.7)$$

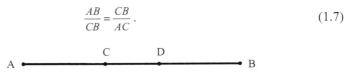

Figure.1.6. The golden ratio

This is the definition of the "golden ratio," used in modern science.

We denote the proportion (1.6) through *x*. Then, taking into consideration, that *AB = AC + CB*, the proportion (1.6) can be written as follows:

$$x = \frac{AC + CB}{CB} = 1 + \frac{AC}{CB} = 1 + \frac{1}{\frac{CB}{AC}} = 1 + \frac{1}{x}. \tag{1.8}$$

The following algebraic equation follows from (1.8):

$$x^2 - x - 1 = 0. \tag{1.9}$$

From the "physical meaning" of the proportion (1.6) implies that we should use the positive root of the equation (1.9). We denote this root through Φ:

$$\Phi = \frac{1 + \sqrt{5}}{2}. \tag{1.10}$$

1.4.2. How did Euclid use the "golden section"?

The question arises: why Euclid introduced in the *Elements* the different forms of the "golden ratio," which we can find in the Books II, VI and XIII? To answer this question, we again return back to the Platonic solids (Fig.1.3). As is known, only three types of regular polygons can be the faces of the Platonic solids: *equilateral triangle* (tetrahedron, octahedron, icosahedron), *square* (cube) and a *regular pentagon* (dodecahedron). In order to construct the Platonic solids, we must first of all be able to build the verge of Platonic solids geometrically (that is, by using a ruler and compass).

(a)   (b)

Figure 1.7. Dodecahedron (a) and icosahedron (b)

Euclid knew, how to construct an *equilateral triangle* and a *square*, but he met some difficulties at constructing of the *regular pentagon*, which underlies the *dodecahedron* (Fig.1.7-a). The *icosahedron* in Fig.7-b is geometric figure dual to the *dodecahedron*.

It is for this purpose, Euclid had introduced in the Book II the "golden ratio" (Proposition II.11), which is presented in the *Elements* in two forms. By using the "golden ratio," Euclid constructs the "golden" isosceles triangle, whose angles at the base are equal to the doubled angle at the vertex (Fig.1.8-a).

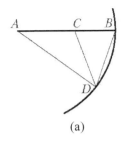 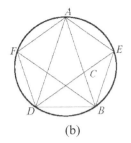

(a)   (b)

Figure 1.8. The "golden" isosceles triangle (a) and the pentagon (b)

At first, the "golden" isosceles triangle is constructed by using a ruler and compass (see Fig.1.8-a). The triangle ABD has the equal sides AB and AD and the equal angles B and D at the base BD. These angles are equal to the doubled angle at the vertex A.

By using the "golden" isosceles triangle ABD (Fig.1.8), we construct the regular pentagon in Fig.1.8-b.

And then we have only one step to the geometric construction of the dodecahedron (Fig 1.7-a), one of the most important regular polyhedra, which symbolized the Universe Harmony in Plato's cosmology.

1.5. The algebraic identities of the golden ratio

1.5.1. The simplest identity

We start with the simplest algebraic properties of the golden ratio. With this purpose we represent the equation (1.9) as follows:

$$x^2 = x+1. \tag{1.11}$$

If we substitute the root Φ (the golden ratio) in place of x into the equation (1.11), then we obtain the first remarkable identity for the golden ratio:

$$\Phi^2 = \Phi + 1. \tag{1.12}$$

If we divide all terms of (1.12) by Φ, we get the next identity for Φ:

$$\Phi = 1 + \frac{1}{\Phi}, \tag{1.13}$$

If we multiply the both parts of (1.12) by the golden ratio Φ, and then divide them by Φ, then we get two new identities:

$$\Phi^3 = \Phi^2 + \Phi \tag{1.14}$$

and

$$\Phi = 1 + \Phi^{-1}. \tag{1.15}$$

If we now continue to multiply the terms of the identity (1.14) by Φ, and to divide the terms of (1.15) by Φ and continue the process to infinity, we obtain the following elegant identity, which connects the degree of the golden ratio:

$$\Phi^n = \Phi^{n-1} + \Phi^{n-2}, \; n = 0, \pm 1, \pm 2, \pm 3, \ldots \tag{1.16}$$

1.5.2. The "golden" geometric progression

Consider the sequence of the powers of the "golden ratio":

$$\left\{ \ldots, \Phi^{-n}, \Phi^{-(n-1)}, \ldots, \Phi^{-2}, \Phi^{-1}, \Phi^0 = 1, \Phi^1, \Phi^2, \ldots, \Phi^{n-1}, \Phi^n, \ldots \right\}. \tag{1.17}$$

The sequence (1.17) has very interesting mathematical properties. On the one hand, the sequence (1.17) is the "geometric progression," in which each term is equal to the previous one, multiplied by the denominator Φ of the geometric progression (1.17), i.e.,

$$\Phi^n = \Phi \times \Phi^{n-1}. \tag{1.18}$$

On the other hand, in accordance to (1.16), the terms of the geometric progression (1.17) have the so-called "summing" property (1.16), because each term of the progression (1.17) is the sum of the two previous terms. Note that the property (1.16) is characteristic only for the geometric progression with the denominator Φ, and therefore such a geometric progression is called the *"golden" progression*.

Since every geometric progression corresponds to some logarithmic spiral, in the opinion of many researchers, the identity (1.16) singles out the "golden" progression (1.17) among other geometric progressions and is a cause of the wide spread of the "golden" progression (1.17) in the forms and structures of Nature.

1.5.3. A representation of the golden ratio in the "radicals"

Let us now consider again the identity (1.12). If you take the square root of the right and left parts of (1.12), we obtain the following expression for Φ:

$$\Phi = \sqrt{1 + \Phi}. \tag{1.19}$$

Now, if we substitute the right-hand side of (1.19) in place of Φ the same expression (1.19) for Φ, we obtain the following expression:

$$\Phi = \sqrt{1+\sqrt{1+\Phi}} . \tag{1.20}$$

If we continue this substitution in the right-hand part of (1.20) an infinite number of times, then we get the following wonderful representation of the golden ratio in the "radicals":

$$\Phi = \sqrt{1+\sqrt{1+\sqrt{1+\sqrt{1+...}}}} . \tag{1.21}$$

1.5.4. A representation of the golden ratio in the form of continued fraction

In order to represent the golden ratio as a continued fraction, we use the identity (1.13). If we substitute into the right-hand part of (1.13) in place of Φ the same expression (1.13), then we came to the representation of Φ in the following form:

$$\Phi = 1 + \cfrac{1}{1+\cfrac{1}{\Phi}} . \tag{1.22}$$

If we continue this substitution in the right-hand part of (1.22) an infinite number of times, then we get the following wonderful representation of the golden ratio in the form of continued fraction:

$$\Phi = 1 + \cfrac{1}{1+\cfrac{1}{1+\cfrac{1}{1+\cfrac{1}{1+...}}}} . \tag{1.23}$$

The expression (1.23) has a deep mathematic sense. The Russian mathematicians A.Y. Khinchin [44] and N.N. Vorobyov [38] drew attention on the fact that the expression (1.23) singles out the golden ratio, among other irrational numbers, because according to (1.23) the golden ratio is approximated by rational numbers the most slowly. This fact emphasizes the uniqueness of the golden ratio among other irrational numbers from the point of view of continued fractions.

We find now suitable fractions for the golden ratio. For this, we use (1.23) to approximate the golden ratio (1.23) with the following suitable fractions:

$$1 = \frac{1}{1} \quad \text{(the first approximation)};$$

$$1 + \frac{1}{1} = \frac{2}{1} \quad \text{(the second approximation)};$$

$$1 + \cfrac{1}{1 + \cfrac{1}{1}} = \frac{3}{2} \quad \text{(the third approximation)};$$

$$1 + \cfrac{1}{1 + \cfrac{1}{1 + \cfrac{1}{1}}} = \frac{5}{3} \quad \text{(the fours approximation)}.$$

By continuing this process, we find a sequence of the suitable continued fractions for the golden ratio, which is a sequence of the adjacent Fibonacci numbers.

$$\frac{1}{1}, \frac{2}{1}, \frac{3}{2}, \frac{5}{3}, \frac{8}{5}, \frac{13}{8}, \frac{21}{13}, \dots \to \Phi = \lim_{n \to +\infty} \frac{n}{n-1} = \frac{1+\sqrt{5}}{2}. \tag{1.24}$$

As is known, the sequence (1.24) expresses the famous law of phyllotaxis [20], according to which Nature constructs pinecones, pineapples, cacti, head of sunflowers and other botanical objects. In other words, Nature uses unique mathematical properties of the golden ratio (1.23), (1.24) in their wonderful creations!

In conclusion, a few words about aesthetic aspects of the identities (1.21) and (1.23). Every mathematician intuitively seeks to express their mathematical results in the simplest, most compact form. And if he finds such "aesthetic form," he gets "aesthetic pleasure." In this respect (the desire for "aesthetic" expression of mathematical results), a mathematical creativity is similar to the creativity of composers or poets, whose main task is to obtain perfect musical or poetic forms, which bring us the "aesthetic pleasure."

Note that the formulas (1.21) and (1.23) are bringing us an "aesthetic pleasure" and an unconscious sense of rhythm and harmony, when we think about the endless repetition of the same simple mathematical elements in the formulas for Φ, given by (1.21), (1.23).

1.6. Geometric properties of the golden ratio and their applications

1.6.1. The "golden" rectangle

A rectangle in Fig.1.9 is called the "golden" rectangle, because the ratio of its larger side to the smaller one is equal to the golden ratio:

$$\frac{AB}{BC} = \Phi = \frac{1+\sqrt{5}}{2}.$$

Consider the golden rectangle with the sides $AB = \Phi$ and $BC = 1$. First we find on the line segments AB and DC such points E and F, which divide the relevant segments AB and DC in the golden ratio. It is clear that $AE=DF=1$, then

$$EB = AB - AE = \Phi - 1 = \frac{1}{\Phi}.$$

Now we draw the line EF, which is called the "*golden line.*" It is clear that the golden line EF divides the golden rectangle $ABCD$ into two new rectangles $AEFD$ and $EBCF$. It follows from the above geometric considerations that the rectangle $AEFD$ is a square.

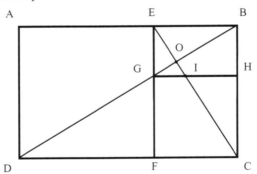

Figure 1.9. The golden rectangle

Consider now the rectangle $EBCF$. Because its larger side $BC=1$ and its smaller side $EB = \frac{1}{\Phi}$, it follows from here that their ratio $\frac{BC}{EB} = \Phi$. Hence, the rectangle $EBCF$ is the "golden" rectangle too! Thus, the golden line EF divides the initial golden rectangle $ABCD$ into the square $AEFD$ and the new golden rectangle $EBCF$.

Draw now the diagonals *DB* and *EC* of the golden rectangles *ABCD* and *EBCF*. It follows from the similarity of the triangles *ABD, FEC, BCE* that the point *G* divides the diagonal *DB* in the golden ratio. Then we draw a new golden line *GH* in the golden rectangle *EBCF*. It is clear that the golden line *GH* divides the golden rectangle *EBCF* into the square *GHCF* and the new golden rectangle *EBHG*. Repeating this procedure infinitely, we get an infinite sequence of squares and golden rectangles converging in the limit to the point *O*.

Note that such infinite repetition of the same geometric figures, that is, squares and golden rectangles, brings us an aesthetic sense of harmony and beauty. It is considered, that this fact is a reason why many objects of rectangular form (match boxes, books, suitcases) frequently have a form of the golden rectangle.

1.6.2. The "golden" brick for the Gothic architecture

Consider the "Euclidean rectangle" in Fig. 1.10, in which the larger side $AB = \Phi$ and the smaller side $AD = \Phi^{-1}$. Draw now the diagonal *DB* in the "Euclidean rectangle" in Fig.1.10.

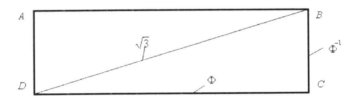

Figure 1.10. Euclidean rectangle

Using "Pythagoras Theorem," we can write:

$$DB^2 = BC^2 + DC^2 = \Phi^2 + \Phi^{-2}. \qquad (1.25)$$

Calculate now the numerical values for Φ^2 and Φ^{-2}. In fact, we have:

$$\Phi^2 = \left(\frac{1+\sqrt{5}}{2}\right)^2 = \frac{1+2\sqrt{5}+5}{4} = \frac{3+\sqrt{5}}{2} \qquad (1.26)$$

$$\Phi^{-2} = \left(\frac{\sqrt{5}-1}{2}\right)^2 = \frac{5-2\sqrt{5}+1}{4} = \frac{3-\sqrt{5}}{2} \qquad (1.27)$$

Taking into consideration (1.26), (1.27), we can write the expression (1.25) as follows:

$$DB^2 = 3 \tag{1.28}$$

It follows from (1.28) that

$$DB = \sqrt{3} \tag{1.29}$$

The "Euclidean rectangle" in Fig.1.10 together with the classical "golden" rectangle in Fig.1.9 can be used for the construction of a special kind of rectangular parallelepiped called a *"golden" brick* (Fig.1.11). Consider this parallelepiped for the case $AB=DC=1$.

Note that the faces of the "golden" brick in Fig.1.11 are the "golden" rectangles in Fig.1.9 and 1.10. The face $ABCD$ is the classical "golden" rectangle with the side ratio Φ. This means that the edge $AD = \Phi^{-1}$. The face $ABGF$ also is the classical "golden" rectangle with the side ratio Φ. This means that the edge $AF = \Phi$. Finally, the rectangle $BCHG$ is the "Euclidean rectangle" with the side ratio Φ^2. This means that the edge $BG = \Phi$ and the edge $BC = \Phi^{-1}$. Note that the diagonal $CG = \sqrt{3}$.

Using the "Pythagoras Theorem," we can calculate the diagonal CF of the "golden" brick:

$$CF = \sqrt{FG^2 + CG^2} = \sqrt{1 + (\sqrt{3})^2} = 2.$$

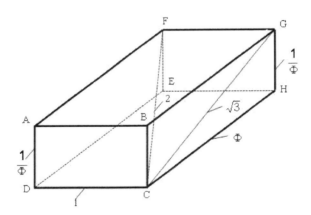

Figure 1.11. The "golden" brick

Jan Grzedzielski in the book [45] draws attention to the fact that the "golden" bricks were widely spread in the Gothic castles as a basic form for the building blocks. There is put forward in [45] the hypothesis that surprising stability of the Gothic castles is connected with the use of the "golden" bricks at the construction of architectural monuments of the gothic style.

1.6.3. The Pentagon and the Pentagram

The golden ratio is a basis of the regular pentagon "(Fig.1.12). The word of "pentagon" is derived from the Greek word of «pentagonon».

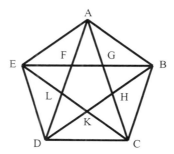

Figure 1.12. The regular pentagon and pentagram

If we draw in a pentagon all diagonals, then we get the *pentagonal star*, also known as a *pentagram* or *pentacle*. The name of "pentagram" is derived from the Greek word «pentagrammon» (pente - five and grammon - line).

It is proved that the points F,G,H,K,L of the intersection of pentagon's diagonals are always the points of the golden ratio. At the same time, they form a new regular pentagon FGHKL. In the new pentagon we can conduct new diagonals. The points of their intersection adduce to the new regular pentagon and this process can be continued indefinitely. Thus, the pentagon ABCDE consists of an infinite number of regular pentagons, which are formed each time by the points of the diagonals intersection. This endless repetition of the same geometric figure (the regular pentagon) creates a sense of rhythm and harmony, which is fixed by our unconscious mind.

The Pythagoreans were delighted with the pentagram, which was considered their main distinctive symbol.

1.6.4. The "golden" cup and the "golden" isosceles triangle

The pentagon and the pentagram in Fig.1.12 include a number of remarkable geometric figures, which had been used widely in the works of arts. In the ancient art is widely known the so-called *"Law of the Golden Cup"* (Fig.1.13), used by the ancient sculptors and goldsmiths. The shaded part of the pentagon in Fig.1.13 gives a schematic representation of the «golden» cup, which resembles the famous Soviet "quality mark."

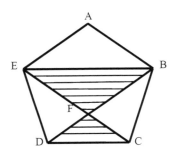

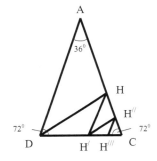

Figure.1.13. The "golden" cup

Figure.1.14. The "golden"
isosceles triangle

The pentagram (Fig.1.12) consists of the five "golden" isosceles triangles (Fig.1.14), each of which resembles the letter "A" ("the five intersecting A"). Each "golden" isosceles triangle in Fig.1.13 has an acute angle $A = 36°$ at the top and two acute angles $D = C = 72°$ at the base of the triangle. The main feature of the "golden" isosceles triangle is that the ratio of each thigh $AC = AD$ to the base DC is equal to the golden ratio Φ.

By exploring the "golden" isosceles triangle (Fig.1.14), as a part of the pentagram (Fig.1.12), the Pythagoreans were delighted when they found that the bisector DH coincides with the diagonal DB of the pentagon (Fig.1.12), and divides the side AC at the point H in the golden ratio (Fig.1.14). As a result, we get the new "golden" triangle DCH.

If we now conduct the bisector of the angle H to the point H' and continue this process indefinitely, we obtain an infinite sequence of the "golden" isosceles triangles. As in the case of the "golden" rectangle (Fig.1.9) and the pentagon

(Fig.1.12), the infinite occurrence of one and the same geometric figure (the "golden" triangle) brings us the aesthetic sense of rhythm and harmony.

1.6.5. Pentacle of the Planet Venus

The Planet Venus describes each eight years absolutely regular pentacle on the big circle of heavenly sphere (Fig.1.15). Ancient people noticed this unique phenomenon and were so surprised, that the Venus and its pentacle became the symbols of perfection and beauty. Today only a few people know, that modern Olympic Games follow the half cycle of Venus. Even less people know that the five-pointed star hardly did not become the symbol of Olympic Games, but this symbol was modified in the latest moment: the five acute ends of the star were replaced with the five rings. In opinion of the organizers, such symbol reflects better a spirit and harmony of the Olympic Games.

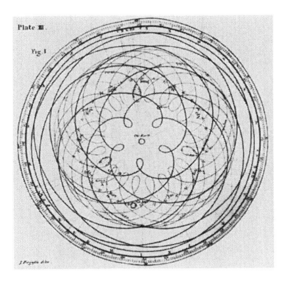

Figure 1.15. Pentacle of the Planet Venus

1.6.6. The golden ratio in the dodecahedron and icosahedron

The dodecahedron (Fig.1.7-a) and the dual to it icosahedron (Fig.1.7-b) take a special place among the Platonic solids. First of all, it must be emphasized that the geometry of the dodecahedron and the icosahedron relate directly to the golden ratio. Indeed, all faces of the dodecahedron are regular pentagons, based on the golden ratio. If we look closely at the icosahedron in Fig.1.7-b, we can see that in each of its vertices the five triangles come together and their outer sides form the regular pentagon. Even these facts are enough to make sure that the golden ratio plays a crucial role in the geometric construction of these Platonic solids.

But there are the deeper mathematical confirmations of the fundamental role of the golden ratio in the icosahedron and the dodecahedron. It is known that the Platonic solids have three specific spheres. The first (inner) sphere is inscribed into the Platonic solid and touches its faces. We denote the radius of the inner sphere by R_i. The second or middle sphere of the Platonic solid touches its ribs. We denote the radius of the middle sphere by R_m. Finally, the third (outer) sphere is described around the Platonic solid and passes through its vertices. We denote its radius by R_c. In geometry, it is proved that the values of the radii of these spheres for the dodecahedron and the icosahedron with an edge of unit length is expressed through the golden ratio (Table 1.1).

Note that the ratio of the radii $\dfrac{R_c}{R_i} = \dfrac{\sqrt{3(3-\Phi)}}{\Phi}$ is the same, for the icosahedron and the dodecahedron. Thus, if the dodecahedron and the icosahedron have the same inner spheres, their outer spheres are equal. This is a reflection of the "hidden harmony" of the dodecahedron and the icosahedron.

Table 1.1. The Golden Ratio in the spheres of the dodecahedron and icosahedron

	R_c	R_m	R_i
Icosahedron	$\dfrac{1}{2}\Phi\sqrt{3-\Phi}$	$\dfrac{1}{2}\Phi$	$\dfrac{\frac{1}{2}\Phi^2}{\sqrt{3}}$
Dodecahedron	$\dfrac{\Phi\sqrt{3}}{2}$	$\dfrac{\Phi^2}{2}$	$\dfrac{\Phi^2}{2\sqrt{3-\Phi}}$

1.6.7. Icosahedron as the main geometrical object of mathematics

In the late of the 19th century, the great German mathematician Felix Klein drew attention to the Platonic solids. He predicted an outstanding role of the Platonic Solids, in particular, the icosahedron for the future development of science and mathematics. In 1884 Felix Klein published the book "Lectures on the Icosahedron" [41], dedicated to the geometric theory of the icosahedron.

Figure 1.16. Felix Klein (1849 - 1925)

According to Klein, the tissue of mathematics widely and freely extends like the sheets of the different mathematical theories. But there are geometric objects, which unite many mathematical theories. Their geometry binds these mathematical theories and allows embracing a general mathematical sense of the miscellaneous theories. In Klein's opinion, the icosahedron is precisely such mathematical object. Klein treats the regular icosahedron as the mathematical object, from which the branches of the five mathematical theories follow, namely, *geometry, Galois' theory, group theory, invariants theory and differential equations.*

Thus, the prominent mathematician Felix Klein after Pythagoras, Plato, Euclid, and Johannes Kepler could estimate a fundamental role of the Platonic Solids, in particular the icosahedron, for the development of science and mathematics. Klein's main idea is extremely simple [36]: *"Each unique geometrical object is somehow or other connected to the properties of the regular icosahedron".*

Unfortunately, Klein's contemporaries could not understand and appreciate a revolutionary importance of Klein's idea, suggested by him in 19th century. However, its significance was appreciated one century after, when the Israeli scientist Dan Shechtman discovered in 1982 a special alloy called *quasi-crystals*

[25] and the outstanding researchers Robert F. Curl, Harold W. Kroto and Richard E. Smalley discovered in 1985 a special kind of carbon called *fullerenes* [24]. It is important to emphasize that the *quasi-crystals* are based on the Platonic icosahedron and the *fullerenes* on the Archimedean truncated icosahedron.

1.6.8. "Parquet's problem" and Penrose's tiling

Since ancient times, there is the "parquet's problem" in geometry, that is, a problem how to fill a plane by regular polygons. Already the Pythagoreans proved that only regular (equilateral) triangles (symmetry axis of the 3-rd order), quadrates (symmetry axis of the 4th order) and hexagons (symmetry axis of the 6th order) can solve this problem. The "parquet's problem" has a direct relation to the main law of a crystallography, according to which only the symmetry axises of the 3-rd, 4-th and 6-th order are allowed in crystals. Symmetry axises of the 5th order and more than 6 are prohibited in crystals.

The English mathematician Roger Penrose was the first scientist, who found other solution of the "parquet's problem". In 1972 he has covered a plane in non-periodic manner, by using only two simple polygons. In the simplest form, "Penrose's tiles" are a nonrandom set of rhombi of two types, the first one (Fig.1.17-a) has the internal angle 72°, and the second one (Fig.1.17-b) has the internal angle 36°.

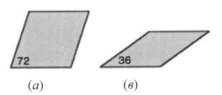

(a)   *(в)*

Figure 1.17. Penrose's rombi

To understand a geometric essence of "Penrose's tiles", we should turn back to the "pentagon" and "pentagram" (Fig.1.12).

As is known, in the pentagram there is a number of the distinctive "golden" isosceles triangles. The triangle of the type *ADC* (Fig.1.14) is the first of them. The acute angle at the vertex of *A* is equal to 36°, and the ratio of the side *AC=AD* to the base *DC* is equal to the golden ratio, that is, the given triangle is the "golden" isosceles triangle. Note that the pentagram consists of 5 equal "golden" isosceles triangles *ADC, BED, CAE, DBA,* and *ECB* similar to Fig.1.14. If we take now such two "golden" triangles and connect them together by their bases, we get "Penrose's tile," represented in Fig.1.17-b and named *"thin"* *rhombus*. The "thin" rhombus has four vertexes with the following angles: 36°, 36°, 144°, 144°.

Besides, we have in the pentagram the other 5 small "golden" isosceles triangles *AGB, AFE, BHC, CKD,* and *DEL.* Let us consider now one more type of the "golden" isosceles triangle, presented in the pentagram, for example, *AGB.* In such "golden" triangle the acute angles at the vertexes *A* and *B* are equal to 72°, and the obtuse angle at the vertex *A* is equal to 108°. Note that the ratio of the base *AB* of the triangle *AGB* to its sides *AE=AB* is equal to the golden ratio, that is, this triangle also is the "golden" isosceles triangle. If we connect now such two triangles together by their bases, we obtain the second "Penrose tile," represented in Fig.1.17-*a* and named *"thick" rhombus.* The thick rhombus has the four vertices with the angles: 72°, 72°, 108°, 108°.

The "thin" and "thick" rhombi in Fig.1.17 allow covering completely an infinite plane. Below in Fig.1.18 we can see a process of sequential construction of "Penrose's tiling." First, we take the 5 "thick" rhombi and connect them together, as shown in Fig. 1.18-*a*. Then, we add to the figure in Fig.1.18-*a* the "thick" and "thin" rhombi, as shown in Fig.1.18-*b*. The Fig.1.18-*c* and 16-*d* are a further development of "Penrose's tiling".

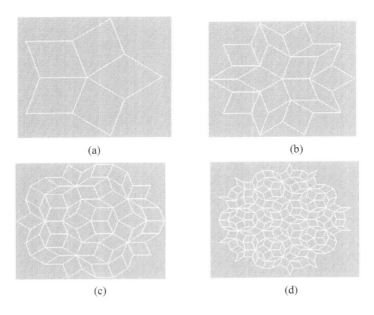

(a)   (b)

(c)   (d)

Figure 1.18. Penrose's tiling

Here there arise some non-periodic structures called *Penrose's tiling*. It is proved that the ratio of the number of the "thick" rhombi (Fig.1.17-*a*) to the number of the "thin" rhombi (Fig.1.17-*b*) in "Penrose's tiling" in Fig.1.18 aims in the limit to the golden ratio.

The mosaic in Fig.1.18 is named after the famous English mathematician and physicist Roger Penrose (Fig.1.19).

Figure 1.19. Sir Roger Penrose stands on the floor, covered with Penrouse's mosaic

1.6.9. Quasi-crystals

On November 12, 1984 in the small article, published in the authoritative journal "Physical Review Letters," the experimental proof of the existence of the metal alloy with exclusive physical properties was presented. The Israeli physicist **Dan Shechtman** was the author of this article.

Figure 1.20. The Israel physicist Dan Shechtman

A special alloy, discovered by Prof. Shechtman in 1982 and called quasi-crystal, is the focus of his research. A crystallographic structure of quasi-crystals is unique, because they have a "pentagonal" symmetry, and their atomic order is quasi-periodic in contrast to periodic order found in previously known crystals. The intriguing questions regarding the structure and properties of this new class of materials have a broad interdisciplinary scientific interest. This field became a major communication arena for physicists, mathematicians, and even representatives of art.

Using methods of electronic diffraction, Shechtman found that the new metallic alloys have all symptoms of crystals. Their diffraction pictures were composed from the bright and regularly located points similar to crystals. However, this picture is characterized by the so-called "icosahedral" or "pentagonal" symmetry, strictly prohibited according to geometric reasons. Such unusual alloys are called "quasi-crystals".

A concept of quasi-crystals generalizes and completes a definition of the crystal. Gratia has written in the article [46]:

"A concept of the quasi-crystals is of fundamental interest, because it extends and completes the definition of the crystal. A theory, based on this concept, replaces the traditional idea about the "structural unit," repeated periodically, onto the key concept of the distant order. This concept resulted in widening crystallography and we only begin to study newly uncovered wealth. Its significance in the world of minerals can be put in one row with adding of the irrational numbers to the rational ones in mathematics".

Shechtman's discovery considerably stimulated researches in this important area. In the recent years, new kinds of the quasi-crystal alloys have been found. It appeared that in addition to the quasi-crystals with the symmetry of the 5-th order, the quasi-crystals with the "decagonal" symmetry (of the 10th order) and the "decagonal" symmetry (of the 12th order) have been discovered.

Penrose's tiling (Fig.1.18) as a "planar analogy" of quasi-crystals was used for the theoretical explanation of the quasi-crystals phenomenon. In the spatial model, the "regular icosahedrons" (Fig.1.7-b) played a role of "Penrose's rhombi" in the planar model. By using the regular icosahedrons, we can fulfill a dense filling of three-dimensional space.

What is practical significance of the quasi-crystals discovery? As Gratia writes [46], *"the mechanical strength of the quasi-crystals increased sharply; here the absence of periodicity resulted in slowing down the distribution of dislocations in comparison to the traditional metals ... This property is of great practical significance: a use of the "icosahedral" phase will allow to get the light and very stable alloys by means of the inclusion of the small-sized fragments of quasi-crystals into aluminum matrix".*

What is the significance of quasi-crystals' discovery from the point of view of the "Mathematics of Harmony"? First of all, this discovery is a confirmation of the great triumph of the "icosahedron-dodecahedron doctrine," which passes through all history of science, starting from Pythagoras, Plato and Euclid, and is a source of the deep and useful scientific ideas. Secondly, the quasi-crystals shattered the conventional presentation about the insuperable watershed between the mineral world, where the "pentagonal" symmetry was prohibited, and the alive world, where the "pentagonal" symmetry is one of most widespread.

And in conclusion it is one more historical remark. Dan Shechtman published his first article about the quasi-crystals in 1984, that is, exactly 100 years later after the publication of Felix Klein's book "Lectures on the Icosahedron ..." (1884) [36]. This means that this discovery is a worthy gift to the centennial anniversary of Klein's book, in which the famous German mathematician Felix Klein predicted the outstanding role of the icosahedron for future development of science.

In 2011, the Israeli physicist Dan Shechtman was awarded the Nobel Prize in Chemistry for the discovery of quasi-crystals.

1.6.10. The truncated icosahedron

Polyhedra, which are derived from the Platonic solids by means of cutting off their tops, are called "truncated polyhedra." In connection with the discovery of fullerenes, awarded in 1996 the Nobel Prize in Chemistry, the truncated icosahedron (Fig.1.21) is the most interesting for modern science. In his Nobel lecture, the American scientist **Richard Smalley**, one of the authors of the experimental discovery of fullerenes, is considering of Archimedes as the first explorer of the "truncated polyhedra."

Figure 1.21. Construction of the Archimedean truncated icosahedron from the Platonic icosahedron

As follows from Fig.1.21, the 5 faces (triangles) converge in each of the 12 vertices of the icosahedron. If we cut off the 12 tops of the icosahedron by a plane, it is formed of the 12 new pentagonal faces. Together with the existing 20-th faces, which turned after this cut-off from triangular to hexagonal, the amount of the faces of the truncated icosahedron will increase to 32 faces. The number of ribs will be 90 and the number of vertices will be 60 (Fig.1.22).

Figure 1.22. The Archimedean truncated icosahedron

1.6.11. The fullerenes

The title of the "fullerene" originates from **Buckminster Fuller** (1895 - 1983), who was the American designer, architect, poet, and inventor. Fuller created a large number of inventions, mostly in the fields of a design and architecture, the best-known of which is the *geodesic dome* (Fig.1.23).

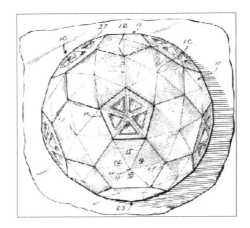

Figure 1.23. Fuller's geodesic dome based on the truncated icosahedron

After the discovery of *fullerenes*, the name of Buckminster Fuller became world-wide famous. The title of "fullerenes" refers to the carbon molecules of the type C_{60}, C_{70}, C_{76}, C_{84}, in which all atoms are on a spherical or spheroid surface. In these molecules, the atoms of carbon are located in vertexes of regular hexagons or pentagons, which cover a surface of sphere or spheroid. We start from a brief history of the molecule C_{60}. This molecule plays a special role among fullerenes. This molecule is characterized by the greatest symmetry and as consequence by the greatest stability. By its shape, the molecule C_{60} reminds a typical white and black football, the Telstar (football) of 1970 (Fig.1.24), which has a structure of the truncated icosahedron (Fig.1.22).

Figure 1.24. The Telstar (football) of 1970

The atoms of carbon in this molecule are located on the spherical surface in the vertexes of 20 regular hexagons and 12 regular pentagons; here each hexagon is connected with three hexagons and three pentagons, and each pentagon is connected with hexagons (Fig.1.25). The most striking property of the C60 molecule is its high symmetry. There are 120 symmetry operations, which convert the molecule onto itself. This makes this molecule the *most symmetric molecule.*

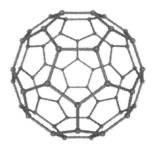

Figure 1.25. A structure of the molecule C_{60}

It is not surprising that the shape of the molecule C_{60} attracted attention of many artists and mathematicians during many centuries. We mentioned above, that the truncated icosahedron was already known to Archimedes. The oldest known picture of the football-shape was found in the Vatican library. This picture was from the book of the painter and mathematician Piero della Francesca. We can find the truncated icosahedron in Pacioli's book "Divina Proportione" (1509). Also Johannes Kepler had studied the Platonic and Archimedean Solids and introduced the name "truncated icosahedron" for this shape.

The *fullerenes,* in essence, are the "man-made" structures, following from fundamental physical researches. They were discovered in 1985 by *Robert F. Curl, Harold W. Kroto* and *Richard E. Smalley*. The researchers named the newly-discovered chemical structure of the carbon C_{60} *buckminsterfullerene* in honor of Buckminster Fuller. In 1996 they got the Nobel Prize in chemistry for this discovery.

The fullerenes possess unusual chemical and physical properties. So, at high pressure the carbon C60 becomes firm, as diamond. Its molecules form the crystal structure as if consisting of ideally smooth spheres, freely rotating in a cubic lattice. Owing to this property, C60 can be used as firm greasing. The fullerenes possess also unique magnetic and superconducting properties.

Chapter 2

FIBONACCI AND LUCAS NUMBERS, BINET'S FORMULAS, AND HILBERT'S TENTH PROBLEM

2.1. A History of the Fibonacci numbers

2.1.1. Leonardo Pisano Fibonacci

The "Middle Ages" in our consciousness associate with the orgies of inquisition, campfires, on which witches and heretics had been burned, and by the Crusades for the "God's Coffin". Science in those times obviously was not "in a center of society attention". In this situation, the appearance of the mathematical book "Liber abaci," written in 1202 by the Italian mathematician Leonardo Pisano (by the nickname of Fibonacci), was an important event in the "scientific life of society". Who was Fibonacci? And why his mathematical works are so important for the West-European mathematics? To answer these questions it is necessary to reproduce the historical epoch, in which Fibonacci lived and worked.

It is necessary to note that the period between the 11th and the 12th centuries was the epoch of brilliant flowering of the Arabian culture, but at the same time the beginning of its downfall. In the end of the 11th century, that is, before the beginning of the Crusades, the Arabs were, doubtlessly, the most educated people in the world and they had surpassed their Christian enemies in this respect. However, already before the Crusades the Arabian influence had penetrated to the West. However the greatest infiltration of the Arabian culture to the West began after the Crusades, which, on the one hand, weakened the Arabian world, but, on the other hand, reinforced the Arabian influence on the Christian West. Not only the Palestinian cotton and sugar, pepper and black wood of Egypt, semi-precious stones and spices of India are sought and estimated by the Christian West in the Arabian world. The West starts to estimate properly the cultural heritage of the "great antique East". The world, discovered by the

West researchers, had blinded them by the art and scientific achievements. In the West world the interest in the Arabian geographical maps, in tutorials on algebra and astronomy, and in Arabian architecture rapidly increased. The emperor Fridrich Gogenstaufen, an apprentice of the Sicilian Arabs and an admirer of the Arabian culture, was one of the most interesting persons of the Crusades epoch, which was a harbinger of the Renaissance. At his palace in Pisa the Great European mathematician of the Middle Ages Leonardo Pisano (by the nickname of Fibonacci that means the son of Bonacci) lived and worked.

Figure 2.1. Fibonacci (about 1170 – after 1228)

We know a little about Fibonacci's life. Even the exact date of his birth is obscure. It is assumed that Fibonacci was born in the eighth decade of the 12th century (presumptively in 1170). His father was a merchant and a government official, a representative of a new class of businessmen generated by the "Commercial Revolution". In that time the city of Pisa was one of the largest commercial Italian centers, which actively cooperated with the Islamic East, and Fibonacci's father traded in one of the trading posts founded by Italians on the northern coast of Africa. Due to this circumstance he could give to his son, the future mathematician Fibonacci, a good mathematical education in one of the Arabian educational institutions.

Moris Cantor, one of the known historians of mathematics, had called Fibonacci *"the brilliant meteor, flashed on the dark background of the West-European Middle Ages."* He supposed that, probably, Fibonacci had perished during one of the Crusades (presumptively in 1228), accompanying the emperor Fridrich Gogenstaufen.

Fibonacci wrote several mathematical works: *"Liber abaci"*, *"Liber quadratorum"*, *"Practica geometriae."* The book "Liber abaci" is the most known of them. This book was published at the life of Fibonacci twice, in 1202 and 1228. Note that Fibonacci conceived the book as the manual for traders, however by its significance this book came out far beyond the trade practice. Fibonacci's book can be considered as mathematical encyclopedia of the Middle Ages. From this point of view the Section 12 is of especial interest. In this Section Fibonacci formulated and solved a number of mathematical problems that proved to be interesting from point of view of general perspectives of mathematics development. This Section constitutes an almost third part of the book and, apparently, Fibonacci drew to it a special attention and showed here the greatest originality.

The task of "rabbits' reproduction" is the most known among the problems, formulated by Fibonacci. This task resulted in the discovery of the numerical sequence 1, 1, 2, 3, 5, 8, 13..., called *Fibonacci numbers*.

2.1.2. An influence of Fibonacci's works on the development of the West-European mathematics

Though Fibonacci was one of the brightest mathematical minds in the history of the West-European mathematics, however his contribution in mathematics had been belittled undeservedly. The Russian mathematician Prof. Vasil'ev in his book "Integer Number" (1919) estimated the significance of Fibonacci's mathematical works for the West-European mathematics as follows:

"The works of the learned merchant from Pisa were so much above the level of mathematical knowledge even of the scientists of that time, that their influence on the mathematical literature becomes noticeable only in two centuries after his death, at the end of the 15th century, when many of his theorems and problems had been entered by Luca Pacioli, professor of many Italian universities and Leonardo da Vinci's friend, in his works and in the beginning of the 16th century, when the group of the talented Italian mathematicians: Ferro, Cardano, Tartalia, Ferrari gave the beginning of the higher algebra thanks to the solution of the cubic and biquadratic equations ".

It follows from this quote, that Fibonacci almost on two centuries surpassed the West-European mathematicians of that time. Fibonacci's historical role for the West-European science is similar to the role of Pythagoras, who got his "scientific education" in the Egyptian and Babylonian science and then had transferred the obtained knowledge to the Greek science. Fibonacci got mathematical education in the Arabian educational institutions. The obtained knowledge, in particular, the Arabian-Hindu decimal notation, was introduced by him to the West-European mathematics. Fibonacci (similarly to Pythagoras) had transferred the Arabian mathematical knowledge into the West-European science through his mathematical works and thanks to this he created the fundamentals for the further development of the West-European mathematics.

2.2. Fibonacci's rabbits

2.2.1. A problem of "rabbit's reproduction"

By irony of his fate, Fibonacci, who gave the outstanding contribution to the development of mathematics, is known in modern mathematics only as the author of interesting numerical sequence called *Fibonacci numbers*. This numerical sequence was obtained by Fibonacci at the solution of the well known *task of "rabbit's reproduction."* The formulation and solution of this problem is considered as Fibonacci's main contribution to the development of combinatorial analysis. Namely with the help of this problem Fibonacci anticipated the recurrence method that is considered now as one of the powerful methods of the combinatorial analysis. The recursive relation, obtained by Fibonacci at the solution of this problem, is considered as the first recursive relation in mathematics history. The essence of the "rabbit's reproduction" problem was formulated by Fibonacci as follows:

"Suppose that in the enclosed place there is one pair of rabbits (female and male) on the first day of January. This pair of rabbits gives birth to a new pair of rabbits on the first day of February and then on the first day of each next month. Each newborn pair of rabbits becomes mature in one month and then gives birth to a new pair of rabbits each month after. There is a question: how many pairs of rabbits will be in the enclosed place in one year, that is, in 12 months since the beginning of reproduction?"

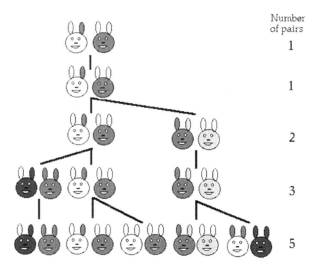

Figure 2.2. Fibonacci's rabbits

For the solution of this problem (Fig.2.2) we designate a pair of the mature rabbits by A, and a pair of the newborn rabbits by B. Then the process of " rabbits' reproduction" can be described with the help of the two "passages" that describes the monthly transformations of the rabbits pairs.

$$A \rightarrow AB \tag{2.1}$$

$$B \rightarrow A \tag{2.2}$$

Note that the passage (2.1) simulates a monthly transformation of each pair of mature rabbits into two pairs, namely, the same pair of the mature rabbits A and the newborn pair B. The passage (2.2) simulates the process of rabbit's "maturing," when the newborn pair B is transformed into the mature pair A. Then, if we begin with the mature pair A, then the process of "rabbit's reproduction" can be represented with the help of Table 2.1.

Table 2.1. Rabbits's reproduction

Data	Pairs of rabbits	A	B	$A+B$
January, 1	A	1	0	1
February,1	AB	1	1	2
March, 1	ABA	2	1	3
April, 1	$ABAAB$	3	2	5
May, 1	$ABAABABA$	5	3	8
June, 1	$ABAABABAABAAB$	8	5	13

Note that in the columns A, B and $A+B$ of Table 2.1 we can see the numbers of mature (A), newborn (B) and total ($A+B$) rabbits, respectively.

2.2.2. Fibonacci's recursive relation

Investigating the A-, B- and ($A+B$)-sequences, it is possible to find the following regularity in these numerical sequences: *each number of the sequence is equal to the sum of two previous.* If now we designate the n-th number of the sequence that satisfies to this rule through F_n, then the above general rule can be written by the following recursive relation:

$$F_n = F_{n-1} + F_{n-2} \tag{2.3}$$

Note that the concrete values of the numeric sequences, generated by the recursive relation (2.3), depend on the initial values (the seeds) of the sequence F_1 и F_2. For example, for the A-numbers we have the seeds:

$$F_1 = F_2 = 1 \tag{2.4}$$

For this case the recursive formula (2.3) at the seeds (2.4) "generates" the following numerical sequence:

$$1,1,2,3,5,8,13,21,34,55,89,144,... \tag{2.5}$$

In mathematics, the numerical sequence (2.5) is called, as a rule, *Fibonacci numbers*. Fibonacci numbers (2.5) have a number of remarkable mathematical properties as is shown below.

2.3. The variations on Fibonacci's theme

The variations on the given theme are a genre well known in music. A distinctive feature of the musical works of the variation genre consists in the fact that they begin, in the most cases, with one simple essential musical theme, which hereinafter undergoes considerable changes by tempo, mood and nature. But how the variations do not bizarre, the listeners absolutely have an impression that each of them is a natural development of the essential theme.

If we follow to the example of musical piece and select a simple mathematical subject (the series of Fibonacci numbers), we can consider this series together with its numerous variations.

2.3.1. The sum of the first n consecutive Fibonacci numbers

Fibonacci numbers have a number of the delightful mathematical properties, which already many centuries shake imagination of mathematicians. Calculate, for example, the sum of the first n Fibonacci numbers. Begin from the simplest sums:

$$\begin{array}{l} 1+1=2=\mathbf{3}-1 \\ 1+1+2=4=\mathbf{5}-1 \\ 1+1+2+3=7=\mathbf{8}-1 \\ 1+1+2+3+5=12=\mathbf{13}-1 \end{array} \tag{2.6}$$

If we consider in the sums (2.6) the numbers marked by bold type: **3, 5, 8, 13,** ..., then it is easy to see, that they are Fibonacci numbers! Then we can write the sums (2.6) as follows:

$$F_1 + F_2 = F_4 - 1; \quad F_1 + F_2 + F_3 = F_5 - 1; \quad F_1 + F_2 + F_3 + F_4 = F_6 - 1;... \tag{2.7}$$

It is clear that the general formula for (2.7) has the following form:

$$F_1 + F_2 + ... + F_n = F_{n+2} - 1. \tag{2.8}$$

2.3.2. The sum of the consecutive Fibonacci numbers with the odd indexes

We start from the simplest sums:

$$\begin{array}{l} 1+2=\mathbf{3} \\ 1+2+5=\mathbf{8} \\ 1+2+5+13=\mathbf{21} \\ 1+2+5+13+34=\mathbf{55} \end{array} \tag{2.9}$$

It follows from (2.9) the following general formula:

$$F_1 + F_3 + F_5 + ... + F_{2n-1} = F_{2n}. \tag{2.10}$$

2.3.3. The sum of the consecutive Fibonacci numbers with the even indexes

$$F_2 + F_4 + F_6 + ... + F_{2n} = F_{2n+1} - 1. \tag{2.11}$$

2.3.4. The sum of the squares of the quadrates of the n consecutive Fibonacci numbers

Find now the sum of the quadrates of the n sequential Fibonacci numbers:

$$F_1^2 + F_2^2 + ... + F_n^2 \tag{2.12}$$

Start from the analysis of the simplest sums of the kind (2.12):

$$\begin{array}{|l|}
\hline
1^2 + 1^2 = 2 = \mathbf{1 \times 2} \\
1^2 + 1^2 + 2^2 = 6 = \mathbf{2 \times 3} \\
1^2 + 1^2 + 2^2 + 3^2 = 15 = \mathbf{3 \times 5} \\
1^2 + 1^2 + 2^2 + 3^2 + 5^2 = 40 = \mathbf{5 \times 8} \\
\hline
\end{array} \tag{2.13}$$

The analysis of (2.13) leads us to the following general formula:

$$F_1^2 + F_2^2 + ... + F_n^2 = F_n F_{n+1} \tag{2.14}$$

2.3.5. The sum of the quadrates of the two adjacent Fibonacci numbers

Let us consider the sum of the quadrates of the two adjacent Fibonacci numbers:

$$F_{n-1}^2 + F_n^2 . \tag{2.15}$$

We start from the analysis of the simplest sums of the kind (2.15):

$$\begin{array}{|l|}
\hline
1^2 + 1^2 = 1 + 1 = \mathbf{2} \\
1^2 + 2^2 = 1 + 4 = \mathbf{5} \\
2^2 + 3^2 = 4 + 9 = \mathbf{13} \\
3^2 + 5^2 = 9 + 25 = \mathbf{34} \\
\hline
\end{array} \tag{2.16}$$

The analysis of (2.16) leads us to the following general formula:

$$F_{n-1}^2 + F_n^2 = F_{2n-1} . \tag{2.17}$$

We take without proof [29] a number of the following remarkable properties of Fibonacci numbers:

$$F_m F_n + F_{m-1} F_{n-1} = F_{m+n-1} \tag{2.18}$$

$$F_{n+1} F_m + F_n F_{m-1} = F_{m+n} . \tag{2.19}$$

In particular, for the case $m = n$ the following formula follows from (2.19):

$$F_{2n} = \left(F_{n-1} + F_{n+1} \right) F_n = \left(2F_{n-1} + F_n \right) F_n . \qquad (2.20)$$

2.3.6. The "extended" Fibonacci numbers

Fibonacci numbers (2.5) can be extended for the negative values of the indexes of n (see Table 2.2)

Table 2.2. The "extended" Fibonacci numbers

n	0	1	2	3	4	5	6	7	8	9	10
F_n	0	1	1	2	3	5	8	13	21	34	55
F_{-n}	0	1	−1	2	−3	5	−8	13	−21	34	−55

The "extended" Fibonacci numbers are connected by the following simple relation:

$$F_{-n} = \left(-1 \right)^{n+1} F_n \qquad (2.21)$$

2.3.7. Cassini's formula for Fibonacci numbers

A history of science keeps in secret why the famous French astronomer **Giovanni Domenico Cassini** (1625-1712) took a great interest in Fibonacci numbers. Most likely it was simply a "hobby" of the great astronomer. At that time many serious scientists took a great interest in Fibonacci numbers and the golden ratio. We can remind, that these mathematical objects were also a hobby of Johannes Kepler, who was Cassini's contemporary.

Consider now the Fibonacci series: $1,1,2,3,5,8,13,21,34,...$ Take the Fibonacci number 5 and to square it, that is, $5^2 = 25$. Now consider the product of the two Fibonacci numbers 3 and 8 that encircle the Fibonacci number 5, that is, $3 \times 8 = 24$. Then we can write:

$$5^2 - 3 \times 8 = 1.$$

Note that the obtained difference is equal to (+1).

And now we will do the same procedure with the next Fibonacci number 8, that is, at first we square it ($8^2 = 64$), then we calculate the product of the two Fibonacci numbers 5 and 13, which encircle the Fibonacci number 8, that is, $5 \times 13 = 65$. After a comparison of the product $5 \times 13 = 65$ to the square $8^2 = 4$ we get:

$$8^2 - 5 \times 13 = -1.$$

Note that the obtained difference is equal to (-1).

Further we have:

$$13^2 - 8 \times 21 = 1;$$

$$21^2 - 13 \times 34 = -1$$

and so on.

We see, that the square of some Fibonacci number F_n always differs from the product of the two adjacent Fibonacci numbers F_{n-1} and F_{n+1}, which encircle it, by 1, but the sign of 1 depends on the index n of the initial Fibonacci number F_n. If the index n is even, then the number 1 is taken with minus, and if odd, with plus. The indicated property of Fibonacci numbers can be expressed by the following mathematical formula:

$$F_n^2 - F_{n-1}F_{n+1} = (-1)^{n+1}. \tag{2.21}$$

This wonderful formula evokes a reverent thrill if to imagine that this one is valid for any value of n (we remind that n can be some integer in limits since $-\infty$ up to $+\infty$), and gives genuine aesthetic enjoying because the alternation of $+1$ and -1 in the expression (2.21) at the successive passing of all Fibonacci numbers produces a feeling of a rhythm and harmony.

2.3.8. Fibonacci numbers in Pascal's triangle

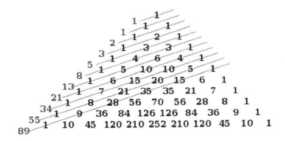

Figure 2.3. Fibonacci numbers in Pascal's triangle

2.4. Lucas numbers

2.4.1. François-Édouard-Anatole Lucas

Fibonacci did not continue to study mathematical properties of the numerical sequence (2.5). This was made by other mathematicians. Since the 19th century the mathematical works devoted to Fibonacci numbers, according to the witty expression of one mathematician, *"began to reproduce as Fibonacci's rabbits."* The French mathematician Lucas became a leader of this study in the 19th century.

Figure 2.4. François-Édouard-Anatole Lucas (1842 – 1891)

The French mathematician **François-Édouard-Anatole Lucas** was born in 1842. He died in 1891 as a result of the accident, which arose at the banquet, when a dish was smashed and a splinter wounded his cheek. Lucas died from infection some days later.

Lucas' major works fall into number theory and indeterminate analysis. In 1878 Lucas gave the criterion for the definition of the primarily of Mersenn's numbers of the kind $M_p = 2^p - 1$. Applying his method, Lucas proved that Mersenn's number

$$M_{127} = 2^{127} - 1 = 170141183460469231731687303715884105727$$

is prime. During 75 years, this number remained the greatest prime number, known in mathematics. Therefore the scientific result, obtained by Lucas in the field of the prime numbers, undoubtedly, belonged to the category of the outstanding mathematical results. Also he found the 12th "perfect number" and formulated a number of interesting mathematical problems.

To honour of Lucas, it is necessary to note scientific prediction in the field of computer science. Already in the 19th century, that is, long before the occurrence of modern computers, Lucas paid attention to technical advantages of the binary notation for the technical realization of computers. It means that he in one century outstripped of the outstanding American physicist and mathematician John von Neumann who gave the preference to the binary numeral system for the technical realization of electronic computers (John von Neumann's principles).

2.4.2. The Lucas numbers

But for our book the most important is the fact that in the 19th century Lucas paid attention of mathematicians on the remarkable numeric sequence (2.5), called by him Fibonacci numbers. Also Lucas had introduced a concept of the generalized Fibonacci numbers, which can be calculated by the following general recurrence relation:

$$G_n = G_{n-1} + G_{n-2},$$ (2.22)

at the arbitrary terms (seeds) G_1 and G_2.

But the main numerical sequence of the type (2.22), introduced by Lucas in the 19th century, is the numerical sequence, which is given by the following recurrence relation:

$$L_n = L_{n-1} + L_{n-2}$$ (2.23)

at the seeds

$$L_1 = 1, L_2 = 3.$$ (2.24)

Then, by using the recurrence relation (2.23) at the seeds (2.24), we can calculate the following numerical sequence called *Lucas numbers*:

$$1, 3, 4, 7, 11, 18, 29, 47, 76, 123, 199, ...$$ (2.25)

If we conduct the reasonings for the Lucas numbers (2.25) similar to the Fibonacci numbers (2.5), we can prove the following identities for Lucas numbers:

$$L_1 + L_3 + L_5 ... + L_{2n-1} = L_{2n} - 2.$$
$$L_1 + L_3 + L_5 ... + L_{2n-1} = L_{2n} - 2$$
$$L_2 + L_4 + L_6 ... + L_{2n} = L_{2n+1} - 1$$

$$L_1^2 + L_2^2 + \ldots + L_n^2 = L_n L_{n+1} - 2$$

$$L_n^2 + L_{n+1}^2 = 5F_{2n+1} \tag{2.26}$$

$$\lim_{n \to \infty} \frac{L_n}{L_{n-1}} = \Phi = \frac{1+\sqrt{5}}{2} \tag{2.27}$$

2.4.3. The "extended" Lucas numbers

Like to the Fibonacci numbers (Table 2.2), the Lucas numbers (2.25) can be extended for the negative values of the indexes n (Table 2.3).

Table 2.3. The "extended" Lucas numbers

n	0	1	2	3	4	5	6	7	8	9	10
L_n	2	1	3	4	7	11	18	29	47	76	123
L_{-n}	2	−1	3	−4	7	−11	18	−29	47	−76	123

The "extended" Lucas numbers are connected by the following simple relation:

$$L_{-n} = (-1)^n L_n \tag{2.28}$$

2.4.5. The remarkable identities for Fibonacci and Lucas numbers

The next group of the formulas is based on the following identity, which connects the generalized Fibonacci numbers G_n (2.22) with the classical Fibonacci numbers F_n [29]:

$$G_{n+m} = F_{m-1}G_n + F_m G_{n+1} \tag{2.29}$$

We can prove this formula by induction on m. For the cases $m=1$ and $m=2$ this formula is valid because

$$G_{n+1} = F_0 G_n + F_1 G_{n+1} = G_{n+1}$$

and

$$G_{n+2} = F_1 G_n + F_2 G_{n+1} = G_n + G_{n+1}$$

The base of the induction is proved.

Suppose now that the formula (2.29) is valid for the cases $m=k$ and $m=k+1$, that is,

$$G_{n+k} = F_{k-1}G_n + F_k G_{n+1}$$

and

$$G_{n+k+1} = F_k G_n + F_{k+1}G_{n+1}.$$

Summing these formulas termwise, we get the following identity:

$$G_{n+k+2} = F_{k+1}G_n + F_{k+2}G_{n+1}$$

The identity (2.29) is proved.

A number of interesting identities for Fibonacci and Lucas numbers follows from the identity (2.29). In particular, for the case $G_i=F_i$ and $m=n +1$ we get from (2.29) the following identity:

$$F_{2n+1} = F_{n+1}^2 + F_n^2, \tag{2.30}$$

what corresponds to the identity (2.17).

2.5. Binet's formulas

2.5.1. Jacques Philippe Marie Binet

Besides François-Édouard-Anatole Lucas, the French mathematician **Jacques Philippe Marie Binet** (1776-1856) was another 19th century enthusiast of Fibonacci numbers and the golden section. He was born on February 2, 1776 in Renje and died on May 12, 1856 in Paris. Binet graduated from the Polytechnic School in Paris and after its graduation in 1806 he worked at the Bridges and Roads Department of the French government. He became a teacher of the Polytechnic school in 1807 and then assistant-professor of the applied analysis and descriptive geometry. Binet investigated the fundamentals of matrix theory and his works in this direction were continued later by other researchers. He found in 1812 the rule of matrix multiplication and already this discovery glorified his name more, than other his works.

Figure 2.5. Jacques Philippe Marie Binet (1776-1856)

Except for mathematics, Binet worked in other areas. He published many articles on mechanics, mathematics and astronomy. In mathematics Binet introduced the notion of the "beta function"; also he considered the linear difference equations with alternating coefficients and established some metric properties of conjugate diameters and so on.

Among different honors, obtained by Binet , it is necessary to mention that he was elected to the Parisian Academy of Sciences in 1843.

2.5.2. Deducing Binet's formulas

Binet came into the Fibonacci numbers theory as the author of the famous mathematical formulas called *Binet's formulas*. These formulas link Fibonacci and Lucas numbers with the golden ratio and, doubtlessly, belong to the category of the most famous mathematical formulas.

In order to deduce Binet's formulas, we consider the remarkable identity connecting the adjacent degrees of the golden ratio Φ:

$$\Phi^n = \Phi^{n-1} + \Phi^{n-2} \ (n = 0, \pm 1, \pm 2, \pm 3, ...) \tag{2.31}$$

By using (2.31), we can write the following expressions for the zero, the first and the minus-first degrees of the golden ratio:

$$\Phi^{-1} = \frac{-1+\sqrt{5}}{2}; \quad \Phi^0 = 1 = \frac{2+0\times\sqrt{5}}{2}; \quad \Phi^1 = \frac{1+\sqrt{5}}{2} \tag{2.32}$$

Using the expressions (2.32) and the identity (2.31), we can represent the second, third and fourth degrees of the golden ratio as follows:

$$\Phi^2 = \Phi^1 + \Phi^0 = \frac{3+\sqrt{5}}{2}; \Phi^3 = \Phi^2 + \Phi^1 = \frac{4+2\sqrt{5}}{2};$$

$$\Phi^4 = \Phi^3 + \Phi^2 = \frac{7+3\sqrt{5}}{2}$$

(2.33)

The analysis of the formulas (2.32), (2.33) allows to see the following general regularity in these formulas: each formula can be represented in the following typical form:

$$\frac{A + B\sqrt{5}}{2}.$$

What are the numerical sequences A and B in the formulas (2.32) and (2.33)? It is easy to see that the numerical sequence A is Lucas numbers: 2, 1, 3, 4, 7, 11, 18, ..., and the numerical sequence B is Fibonacci numbers: 0, 1, 1, 2, 3, 3, 5, 8, ... It follows from this reasoning that the general formula, which allows representing the n-th degree of the golden ratio through Fibonacci and Lucas numbers, has the following form:

$$\Phi^n = \frac{L_n + F_n\sqrt{5}}{2}.$$

(2.34)

Note that the formula (2.34) is valid for each integer n taking the values from the set $(n = 0, \pm1, \pm2, \pm3, ...)$.

By using the formula (2.34), we can represent the "extended" Fibonacci and Lucas numbers through the golden ratio. For this purpose it is enough to write the formulas for the sum and the difference of the n-th degrees of the golden ratio $\Phi^n + \Phi^{-n}$ and $\Phi^n - \Phi^{-n}$ as follows:

$$\Phi^n + \Phi^{-n} = \frac{(L_n + L_{-n}) + (F_n + F_{-n})\sqrt{5}}{2}$$

(2.35)

$$\Phi^n - \Phi^{-n} = \frac{(L_n - L_{-n}) + (F_n - F_{-n})\sqrt{5}}{2}$$

(2.36)

Consider now the expressions of the formulas (2.35) and (2.36) for the even values of the indexes $n=2k$. For this purpose, we can recall the property (2.21) of the "extended" Fibonacci numbers (Table 2.2) and the property (2.28) of the "extended" Lucas numbers (Table 2.3). Then for the case $n = 2k$ the formulas (2.35) and (2.36) take the following simple form:

$$\Phi^{2k} + \Phi^{-2k} = L_{2k}$$

(2.37)

$$\Phi^{2k} - \Phi^{-2k} = F_{2k}\sqrt{5} \qquad (2.38)$$

For the odd indexes $n=2k+1$ by using the properties (2.21), (2.28), we can represent the formulas (2.35), (2.36) as follows:

$$\Phi^{2k+1} + \Phi^{-(2k+1)} = F_{2k+1}\sqrt{5} \qquad (2.39)$$

$$\Phi^{2k+1} - \Phi^{-(2k+1)} = L_{2k+1}. \qquad (2.40)$$

By using the formulas (2.37) – (2.40), we can represent the "extended" Fibonacci and Lucas numbers in the following compact form:

$$L_n = \begin{cases} \Phi^n + \Phi^{-n} & \text{for } n = 2k \\ \Phi^n - \Phi^{-n} & \text{for } n = 2k+1 \end{cases} \qquad (2.41)$$

$$F_n = \begin{cases} \dfrac{\Phi^n + \Phi^{-n}}{\sqrt{5}} & \text{for } n = 2k+1 \\ \dfrac{\Phi^n - \Phi^{-n}}{\sqrt{5}} & \text{for } n = 2k \end{cases} \qquad (2.42)$$

The analysis of the formulas (2.41), (2.42) gives us a possibility to feel "aesthetic pleasure" and once again to be convinced in the power of mathematics! Really, we know that the "extended" Fibonacci and Lucas numbers always are integers. On the other hand, any degree of the golden ratio is irrational number. It follows from here that the integer numbers L_n and F_n can be represented with the help of the formulas (2.41) and (2.42) through a special irrational number, the golden ratio Φ!

2.5.3. The examples

For the example, according to (2.41), (2.42) we can represent the Lucas number 3 ($n=2$) and the Fibonacci numbers 5 ($n=5$) as follows:

$$3 = \left(\frac{1+\sqrt{5}}{2}\right)^2 + \left(\frac{1+\sqrt{5}}{2}\right)^{-2}, \qquad (2.43)$$

$$5 = \frac{\left(\dfrac{1+\sqrt{5}}{2}\right)^5 + \left(\dfrac{1+\sqrt{5}}{2}\right)^{-5}}{\sqrt{5}}. \qquad (2.44)$$

It is easy to prove that the identity (2.43) is valid because according to (2.34) we have:

$$\begin{cases} \left(\dfrac{1+\sqrt{5}}{2}\right)^{2} = \dfrac{L_{2}+F_{2}\sqrt{5}}{2} = \dfrac{3+\sqrt{5}}{2}; \\ \left(\dfrac{1+\sqrt{5}}{2}\right)^{-2} = \dfrac{L_{-2}+F_{-2}\sqrt{5}}{2} = \dfrac{3-\sqrt{5}}{2} \end{cases} \qquad (2.45)$$

By using (2.45), we can rewrite the identity (2.43) as follows:

$$3 = \frac{3+\sqrt{5}}{2} + \frac{3-\sqrt{5}}{2}. \qquad (2.46)$$

We can be convinced in the validity of the identity (2.44), if we recall that according to (2.34) we have the following representations:

$$\begin{cases} \left(\dfrac{1+\sqrt{5}}{2}\right)^{5} = \dfrac{L_{5}+F_{5}\sqrt{5}}{2} = \dfrac{11+5\sqrt{5}}{2} \\ \left(\dfrac{1+\sqrt{5}}{2}\right)^{-5} = \dfrac{L_{-5}+F_{-5}\sqrt{5}}{2} = \dfrac{-11+5\sqrt{5}}{2} \end{cases} \qquad (2.47)$$

By using (2.45), we can rewrite the identity (2.44) as follows:

$$5 = \frac{11+5\sqrt{5}}{\sqrt{5}} + \frac{-11+5\sqrt{5}}{\sqrt{5}} = \frac{10\sqrt{5}}{2\sqrt{5}},$$

from where it follows a validity of the identity (2.44).

Note that this reasoning is of a general character, that is, for any "extended" Lucas or Fibonacci numbers, given by the formulas (2.41) and (2.42), all "irrationalities" in the right-hand parts of (2.41) and (2.42) are annihilated mutually always and we get integer numbers as an outcome!

2.5.4. About one historical analogy

A situation with the mutual annihilation of all "irrationalities" in the formulas (2.41), (2.42) reminds the situation, which arose in mathematics for the case of the introduction of complex numbers. In the 16th century the Italian mathematicians contributed essentially to the development of algebra: they can solve in radicals the equation of 3-rd and 4-th degrees. The famous Cardano's book *The Great Art* (1545) contained the algebraic solution of the cubic equation:

$$x^3 + px + q = 0 \qquad (2.48)$$

in the following form:

$$x = u + v,$$

where

$$\begin{cases} u = \sqrt[3]{-\dfrac{q}{2} + \sqrt{\left(\dfrac{q}{2}\right)^2 + \left(\dfrac{p}{3}\right)^3}} \\[4mm] v = \sqrt[3]{-\dfrac{q}{2} - \sqrt{\left(\dfrac{q}{2}\right)^2 + \left(\dfrac{p}{3}\right)^3}} \\[4mm] uv = -\dfrac{p}{3} \end{cases}$$

It was proved in algebra that the algebraic equation (2.48) has three roots:

(1) For the case $\Delta = \sqrt{\left(\dfrac{q}{2}\right)^2 + \left(\dfrac{p}{3}\right)^3} > 0$, Eq. (2.48) has one real root and two complex conjugate roots; for example, the equation $x^3 + 15x + 124 = 0$ with $\Delta > 0$ has the following roots: $x_1 = -4; x_{2,3} = 2 \pm 3i\sqrt{3}$.

(2) For the case $\Delta = 0$, $p \neq 0$, $q \neq 0$, Eq. (2.48) has three real roots; for example, the roots of the equation $x^3 - 12x + 6 = 0$ are the following: $x_1 = -4; x_{2,3} = -2$.

(3) For the case $\Delta < 0$, we have the most interesting case, the so-called "non-reducible" case, when we need to extract the root of the 3-d degree from complex numbers and the cubic roots u and v are complex numbers. Nevertheless, in this case the equation (2.48) has real roots. For example, the equation $x^3 - 21x + 20 = 0$ with

$$\Delta = -243, \ u = \sqrt[3]{-10 + \sqrt{-243}}, \ v = \sqrt[3]{-10 - \sqrt{-243}} \qquad (2.49)$$

has the real roots 1, 4, -5!

This fact seemed paradoxical for the 16th century mathematicians! Really, all factors of the equation $x^3 - 21x + 20 = 0$ are real numbers, all its roots are real numbers, but the intermediate calculations leads to the "imaginary," "false," "nonexistent" numbers of the kind (2.49). Mathematicians turned out in very difficult situation what happened with them repeatedly (starting from the discovery of irrationals). Completely to ignore the numbers of the kind (2.49) would mean to refuse the general formulas for the solution of the algebraic equations of the 3-rd degree, and also other remarkable mathematical achievements. On the other hand, to recognize that these obtrusively appearing "monstrous" numbers such as (2.49) are equivalent with real numbers was inadmissible from the point of view of common sense. A long time the "monstrous" numbers of the kind (2.49) were not recognized by many mathematicians. For example, Descartes considered that complex numbers do not have any real interpretation and that they are doomed forever to remain only "imaginary" numbers (the name "imaginary numbers" came into mathematics in 17th century after Descartes). Many great mathematicians, in particular, Newton and Leibniz, adhered to the same opinion.

In the 17th century one the English mathematician Vallis in his book *Algebra: historical and practical treatise* (1685) had pointed out a possibility of the geometric interpretation of complex numbers. Finally, the complex numbers came in the use in the 18th century after the works of the French mathematician Moivre (1667-1754) who introduced the following well known formula:

$$\left(\cos\varphi \pm i\sin\varphi\right)^n = \cos n\varphi \pm i\sin n\varphi, \quad i = \sqrt{-1} \; . \tag{2.50}$$

After the introduction of Moivre's formula (2.50), the representation of complex numbers in trigonometric form came in the use what facilitated the solution of many mathematical problems. However, the famous "Euler's formulas" became by the "celebration moment" for the complex numbers. Using Moivre's formula (2.50), Euler proved the following formulas for trigonometric functions:

$$\cos x = \frac{e^{xi} + e^{-xi}}{2}, \quad \sin x = \frac{e^{xi} - e^{-xi}}{2i} \; . \tag{2.51}$$

Note that finding the connection between trigonometric and exponential functions, expressed by Euler's formulas, emphasizes a fundamental connection between the numbers π and e, two numerical constants of mathematics, what played a great role in mathematics, in particular, in the development of the complex number concept.

Returning back to Binet's formulas (2.41) and (2.42) and taking into consideration our reasoning, concerning complex numbers, we can put forward a supposition, that Binet's formulas touch upon some rather deep number-theoretical problems, which are on the intersection of integers (Fibonacci and Lucas numbers) and irrationals (the golden ratio).

2.6. A discovery of the American wunderkind George Bergman

2.6.1. Number system with irrational base

In 1957 the American wunderkind **George Bergman** published the paper *A number system with an irrational base* [47] in the authoritative journal *"Mathematics Magazine."* Possibly, the numeral system with irrational base, developed by George Bergman in 1957, is the most important mathematical discovery in the field of numeral systems after the discovery of positional principle of number representation (Babylon, 2000 B.C.) and decimal numeral system (India, 5th century). However, the most surprising is the fact that George Bergman made his mathematical discovery in the age of 12 years! This is an unprecedented event in the history of mathematics!

The following sum is called *Bergman's number system*:

$$A = \sum a_i \Phi^i \ (i = 0, \pm 1, \pm 2, \pm 3, ...),\tag{2.52}$$

where A is a real number, a_i is a binary numeral (0 or 1) of the i-th digit, Φ^i is the weight of the i-th digit, Φ is the base of number system (2.52).

Let us consider the binary system

$$A = \sum a_i 2^i \ (i = 0, \pm 1, \pm 2, \pm 3, ...),\tag{2.53}$$

which underlies modern computer technology

At the first sight, there is not any distinction between the formulas (2.52) and (2.53) but it is only at the first sight. A principal distinction of the number system (2.52) from the binary system (2.53) is the fact that the irrational number $\Phi = \frac{1+\sqrt{5}}{2}$ (the golden ratio) is used as a radix of the number system (2.52). That is why, Bergman called it the *"number system with an irrational base."* Although Bergman's article [47] contained the result of principal importance for the numeral system theory, however in that period this article simply did not be noted neither by mathematicians nor by engineers. And in conclusion of his article [47] George Bergman wrote: *"I do not know of any useful application for systems such as this, except as a mental exercise and pastime, though it may be of some service in algebraic number theory."*

2.6.2. A representation of the powers of the golden ratio

The 0th power of the golden ratio is represented as follows:

$$\Phi^0 = 1 = 1.0 . \tag{2.54}$$

The positive and negative powers of the golden ratio are represented as follows:

$$\begin{cases} \Phi^1 = 10; & \Phi^{-1} = 0.1; \\ \Phi^2 = 100; & \Phi^{-2} = 0.01; \\ \Phi^3 = 1000; & \Phi^{-3} = 0.001. \end{cases} \tag{2.55}$$

Consider now the "golden" representations of the sums of the golden ratio powers. For example, the sum

$$A = \Phi^4 + \Phi^3 + \Phi^0 + \Phi^{-1} + \Phi^{-2} + \Phi^{-5} \tag{2.56}$$

has the following "golden" representation:

$$A = 11001.11001 . \tag{2.57}$$

By using the formula (2.34), we can rewrite the sum (2.56) as follows:

$$\begin{aligned} A = & \frac{L_4 + F_4\sqrt{5}}{2} + \frac{L_3 + F_3\sqrt{5}}{2} + \frac{L_0 + F_0\sqrt{5}}{2} + \frac{L_{-1} + F_{-1}\sqrt{5}}{2} \\ & + \frac{L_{-2} + F_{-2}\sqrt{5}}{2} + \frac{L_{-5} + F_{-5}\sqrt{5}}{2} \end{aligned} \tag{2.58}$$

Substituting the values of the Fibonacci and Lucas numbers, taken from Tables 2.2, 2.3:

$$L_4 = 7; L_3 = 4; L_0 = 2; L_{-1} = -1; L_{-2} = 3; L_{-5} = -11;$$
$$F_4 = 3; F_3 = 2; F_0 = 0; F_{-1} = 1; F_{-2} = -1; F_{-5} = 5$$

into the expression (2.58) we get the number A in the explicit form:

$$A = \frac{4 + 10\sqrt{5}}{2} = 2 + 5\sqrt{5} . \tag{2.59}$$

Notice that all real numbers, given by the expressions (2.55), (2.57), and the number A, given by (2.56), (2.59), are irrational numbers! But according to (2.55), (2.57) all they are represented by finite combinations of binary numerals. This means that Bergman system (2.52) allows to represent some irrational numbers (in particular, the powers of the golden ratio and their sums) by the finite combinations of binary numerals what is theoretically impossible for the binary system (2.53). This result is the first unusual property of Bergman's system (2.52).

2.6.3. The "golden" representations of natural numbers

Let us consider the "golden" representation of natural numbers in the form (2.52), that is,

$$N = \sum_i a_i \Phi^i . \tag{2.60}$$

We name the sum (2.60) Φ-*code of the natural number N*. The abridged notation of the Φ-code (2.60) of the natural number N has the following form:

$$N = a_n a_{n-1}...a_1 a_0 . a_{-1} a_{-2}...a_{-k} \tag{2.61}$$

Here, the point (after the numeral a_0 divides the "golden" representation (2.61) into two parts: the left-hand part of the digits $a_n a_{n-1}...a_1 a_0$, corresponding to the weights $\Phi^n, \Phi^{n-1},..., \Phi^1, \Phi^0$, and the right-hand part of the digits $a_{-1} a_{-2}...a_{-k}$, corresponding to the weights $\Phi^{-1}, \Phi^{-2},..., \Phi^{-k}$.

To get the "golden" representations of all natural numbers, we use the rule

$$N' = N+1. \tag{2.62}$$

To apply the rule (2.62) to get the "golden" representation of the number N' from the "golden" representation of the previous number N, we need to transform the "golden" representation (2.61) of the initial number N to such form when the binary numeral of the 0-th digit is equal to 0, i.e. $a_0 = 0$. We can fulfill such transformation by means of the micro operations of "convolution" and "devolution," based on the fundamental property (2.31):

$$\underline{\text{Convolution:}} \ 011 \to 100 \left(\Phi^i + \Phi^{i-1} = \Phi^{i+1} \right) \tag{2.63}$$

$$\underline{\text{Devolution:}} \ 100 \to 011 \left(\Phi^{i+1} = \Phi^i + \Phi^{i-1} \right) \tag{2.64}$$

If we add the binary 1 to the 0-th digit of the "golden" representation (2.61), we can realize the rule (2.62).

Demonstrate this method for the case of the "golden" representation of the number 1:

$$1 = 1.0 \tag{2.65}$$

By using the micro operation of the "devolution" (100=011), we get another "golden" representation of the number 1:

$$1 = \Phi^0 = 1.0 = 0.11 = \Phi^{-1} + \Phi^{-2} \qquad (2.66)$$

Then we apply the rule (2.62) to the "golden" representation (2.66). To fulfill this transformation we should add the bit 1 to the 0-th digit of the "golden" representation (2.66). As result, we get the "golden" representation of the number 2:

$$2 = \Phi^0 + \Phi^{-1} + \Phi^{-2} = 1.11 \qquad (2.67)$$

If we fulfill the operation of the "convolution" in the "golden" representation (2.67), we get another "golden" representation of the number 2:

$$2 = 10.01 = \Phi^1 + \Phi^{-2} \qquad (2.68)$$

Adding the bit 1 to the 0-th digit of the "golden" representation (2.68) and performing the "convolution," we get the following "golden" representation of the number 3:

$$3 = 11.01 = 100.01 = \Phi^2 + \Phi^{-2}. \qquad (2.69)$$

The "golden" representation of the number 4 is the following:

$$4 = 101.01 = \Phi^2 + \Phi^0 + \Phi^{-2}. \qquad (2.70)$$

We can get the "golden" representation of the number 5 from the representation (2.70) after the following transformations of the "golden" representation of the number 4, based on the "devolution":

$$4 = 101.01 = 101.0011 = 100.1111. \qquad (2.71)$$

Adding the bit 1 to the 0-th digit of the "golden" representation (2.71) and fulfilling all micro operations of the "convolutions," we get the following "golden" representation of the number 5:

$$5 = 101.1111 = 110.0111 = 1000.1001 = \Phi^3 + \Phi^{-1} + \Phi^{-4} \qquad (2.72)$$

Continuing this process, we can get the "golden" representations of all natural numbers. Thus, this study leads us to the following unexpected result, which can be formulated in the form of the following theorem.

Theorem 2.1. All natural numbers can be represented in Bergman's system (2.52) by the finite number of bits.

This means that every natural number N can be represented as a finite sum of the degrees of the golden ratio. Since all the degrees of the golden ratio (except $\Phi^0 = 1$) are irrational numbers such as $\Phi^i (i = \pm1, \pm2, \pm3, ...)$, this statement is far from obvious.

2.6.4. *Z*- and *D*-properties of natural numbers

As is proved in [42], Bergman's system (2.52) is a source for new number-theoretical results. The *Z*- and *D*-properties of natural numbers belong to such results. These properties follow from the following very simple reasoning.

Consider now the representation of the natural number N in Bergman's system, given by (2.60). The representation of the natural number N in the form (2.60) is called *the Φ-code of natural number N* [42].

Note that according to Theorem 2.1, the sum (2.60) is a finite sum for the arbitrary natural number N.

If we use the formula (2.34), we can represent the Φ-code of N (2.60) as follows:

$$N = \frac{1}{2}\left(A + B\sqrt{5}\right) \tag{2.73}$$

where

$$A = \sum_i a_i L_i \tag{2.74}$$

$$B = \sum_i a_i F_i \tag{2.75}$$

where $a_i (i = 0, \pm1, \pm2, \pm3, ...)$ is the ith binary numeral of the Φ-code (2.60), F_i, L_i are the "extended" Fibonacci and Lucas numbers, respectively.

Represent now the expression (2.73) as follows:

$$2N = A + B\sqrt{5} \tag{2.76}$$

Note that the expression (2.76) has a general character and is valid for the arbitrary natural number N.

Let us analyze the expression (2.76). It is clear that the number $2N$, standing in the left-hand part of the expression (2.76), is an even number always. The right-hand part of the expression (2.76) is the sum of the number A and the product of the number B by the irrational number $\sqrt{5}$. But according to (2.74) and (2.75), the numbers A and B are integers always, because the "extended" Fibonacci and Lucas numbers are integers (see Tables 2.2, 2.3). Then, it follows from (2.76) that for the given natural number N the even number $2N$ is equal identically to the sum of the integer A and the product of the integer B by the irrational number $\sqrt{5}$. And this statement is valid for all natural number N! Then there is the question: for what condition the identity (2.76) could be valid in general case? The answer to this question is very simple: the identity (2.76) can be valid for the arbitrary natural number N only if the sum (2.75) is equal to 0 ("zero") identically and the sum (2.74) is equal to the double number of N, that is

$$B = \sum_i a_i F_i \equiv 0 \tag{2.77}$$

$$A = \sum_i a_i L_i \equiv 2N . \tag{2.78}$$

Compare now the sums (2.75) and (2.60). Because the binary numerals a_i $(i = 0, \pm 1, \pm 2, \pm 3, ...)$ in these sums coincide, this means that the expression (2.75) can be obtained from the expression (2.60), if we substitute the "extended" Fibonacci number F_i $(i = 0, \pm 1, \pm 2, \pm 3, ...)$ in place of the degree of the golden ratio Φ^i in the expression (2.60). But according to (2.77) the sum (2.75) is equal identically to 0 independently on the initial natural number N in the expression (2.60). Thus, we have found a fundamentally new property of natural numbers, which can be formulated as follows.

Theorem 2.2 (Z-property of natural numbers). If we represent the arbitrary natural number N in Bergman's system (2.60) and then substitute the "extended" Fibonacci number F_i $(i = 0, \pm 1, \pm 2, \pm 3, ...)$ in place of the degree of the golden ratio Φ^i, then the sum, arising as a result of such a substitution, is equal to 0 identically independently on the initial natural number N, that is, $\sum_i a_i F_i \equiv 0$.

Compare now the sums (2.74) and (2.60). Because the binary numerals a_i $(i = 0, \pm 1, \pm 2, \pm 3, ...)$ in these sums coincide, this means that the expression (2.74) can be obtained from the expression (2.60) if we substitute the "extended" Lucas number L_i $(i = 0, \pm 1, \pm 2, \pm 3, ...)$ in place of the degree of the golden ratio Φ^i in the expression (2.60). But according to (2.78), the sum (2.74) is equal identically to $2N$ independently on the initial natural number N in the expression (2.60). Thus, we have found a fundamentally new property of natural numbers that can be formulated as follows.

Theorem 2.3 (*D*-property). If we represent the arbitrary natural number N in Bergman's system (2.60) and then substitute the "extended" Lucas number L_i $(i = 0, \pm1, \pm2, \pm3, ...)$ in place of the degree of the golden ratio Φ^i in the expression (2.60), then the sum, arising as a result of such substitution is equal to $2N$ identically independently on the initial natural number N, that is, $\sum_i a_i L_i \equiv 2N$.

2.6.5. The example

As example we consider the "golden" representation of the decimal number 10 in Bergman's system:

$$10 = 10100.0101 = \Phi^4 + \Phi^2 + \Phi^{-2} + \Phi^{-4} . \tag{2.79}$$

By using the formula (2.34), we can represent the sum (2.79) as follows:

$$10 = \Phi^4 + \Phi^2 + \Phi^{-2} + \Phi^{-4} = \frac{L_4 + F_4\sqrt{5}}{2} + \frac{L_2 + F_2\sqrt{5}}{2} +$$
$$\frac{L_{-2} + F_{-2}\sqrt{5}}{2} + \frac{L_{-4} + F_{-4}\sqrt{5}}{2} \tag{2.80}$$

If we take into consideration the relations (2.21) and (2.28), which connect the "extended" Fibonacci and Lucas numbers ,

$$L_{-2} = L_2; L_{-4} = L_4; F_{-2} = F_2; F_{-4} = F_4$$

we get from the expression (2.80) the following result:

$$10 = \frac{2(L_4 + L_2)}{2} = L_4 + L_2 = 7 + 3 .$$

Also we can check the sum (2.79) by the Z- and D-properties. If we substitute in (2.79) the "extended" Fibonacci and Lucas numbers F_i and L_i in place of all digress Φ^i we get the following sums:

$$F_4 + F_2 + F_{-2} + F_{-4} = 3 + 1 + (-1) + (-3) \equiv 0 \text{ (Z-property)}$$

$$L_4 + L_2 + L_{-2} + L_{-4} = 7 + 3 + 3 + 7 \equiv 20 = 2 \times 10 \text{ (D-property)}$$

Assessing the general mathematical significance of the Z- and D-properties, given by the theorems 2.2, 2.3, it is necessary to emphasize that these mathematical results are valid only for natural numbers. This means that a new fundamental properties of natural numbers are discovered in the 21th century in the "mathematics of harmony" [19], after 2.5 thousand years from the beginning of their theoretical study.

These mathematical results are the beginning of the new number theory, which is called the *"golden" number theory* [42] and is based on the "golden ratio." The "golden" number theory dates back to the mathematical discovery of the American wunderkind George Bergman [47] and overturns our ideas about numeral systems, moreover, our understanding about relationship between rational and irrational numbers.

2.7. Fibonacci *Q*-matrices

2.7.1. Definition

In the last decades the theory of the Fibonacci numbers [38-40] was supplemented by the theory of the so-called *Fibonacci Q-matrix* [39]. The latter is the 2×2 matrix of the following form:

$$Q = \begin{pmatrix} 1 & 1 \\ 1 & 0 \end{pmatrix}. \tag{2.81}$$

Note that the determinant of the *Q*-matrix (2.81) is equal:

$$\det Q = -1. \tag{2.82}$$

In the paper [48], devoted to the memory of Verner E. Hoggat, the founder of the Fibonacci Association, it was set forth the history of the *Q*-matrix, it was given an extensive bibliography on the *Q*-matrix and related questions and it was emphasized Hoggatt's contribution in the development of the *Q*-matrix theory.

Although the name of the *"Q-matrix"* was introduced before Verner E. Hoggat, just from Hoggatt's papers the idea of the *Q*-matrix *"caught on like wildfire among Fibonacci enthusiasts. Numerous papers have appeared in "The Fibonacci Quarterly" authored by Hoggatt and/or his students and other collaborators where the Q-matrix method became a central tool in the analysis of Fibonacci properties»* [48].

2.7.2. The properties of the Fibonacci Q-matrices

Let us consider here a theory of the Q-matrix, developed in Hoggatt's book [39].

The following theorem, proved in [39], gives a connection between the Q-matrix (2.81) and Fibonacci numbers.

Theorem 2.4. For a given integer n the nth degree of the Q-matrix is given by

$$Q^n = \begin{pmatrix} F_{n+1} & F_n \\ F_n & F_{n-1} \end{pmatrix}, \tag{2.83}$$

where F_{n-1}, F_n, F_{n+1} are the "extended" Fibonacci numbers.

The next theorem [39] gives a formula for the determinant of the matrix (2.83).

Theorem 2.5. For a given integer n we have:

$$\det Q^n = (-1)^n. \tag{2.84}$$

The following remarkable property for the "extended" Fibonacci numbers follows from Theorem 2.5:

$$\det Q^n = F_{n-1}F_{n+1} - F_n^2 = (-1)^n. \tag{2.85}$$

Remind that the identity (2.85) is one of the most important identities for Fibonacci numbers. This one is called *Cassini's formula* after the famous French astronomer **Giovanni Domenico Cassini** (1625-1712).

Represent now the matrix (2.83) as follows:

$$\begin{aligned} Q^n &= \begin{pmatrix} F_n + F_{n-1} & F_{n-1} + F_{n-2} \\ F_{n-1} + F_{n-2} & F_{n-2} + F_{n-3} \end{pmatrix} \\ &= \begin{pmatrix} F_n & F_{n-1} \\ F_{n-1} & F_{n-2} \end{pmatrix} + \begin{pmatrix} F_{n-1} & F_{n-2} \\ F_{n-2} & F_{n-3} \end{pmatrix} = Q^{n-1} + Q^{n-2} \end{aligned} \tag{2.86}$$

Also we can represent the expression (2.86) as follows:

$$Q^{n-2} = Q^n - Q^{n-1} \tag{2.87}$$

By using the recurrent relations (2.86) and (2.87), we can represent the matrices Q^n and Q^{-n} in the explicit form as is shown in Table 2.4.

Table 2.4. The Q-matrices

n	0	1	2	3	4	5
Q^n	$\begin{pmatrix} 1 & 0 \\ 0 & 1 \end{pmatrix}$	$\begin{pmatrix} 1 & 1 \\ 1 & 0 \end{pmatrix}$	$\begin{pmatrix} 2 & 1 \\ 1 & 1 \end{pmatrix}$	$\begin{pmatrix} 3 & 2 \\ 2 & 1 \end{pmatrix}$	$\begin{pmatrix} 5 & 3 \\ 3 & 2 \end{pmatrix}$	$\begin{pmatrix} 8 & 5 \\ 5 & 3 \end{pmatrix}$
Q^{-n}	$\begin{pmatrix} 1 & 0 \\ 0 & 1 \end{pmatrix}$	$\begin{pmatrix} 0 & 1 \\ 1 & -1 \end{pmatrix}$	$\begin{pmatrix} 1 & -1 \\ -1 & 2 \end{pmatrix}$	$\begin{pmatrix} -1 & 2 \\ 2 & -3 \end{pmatrix}$	$\begin{pmatrix} 2 & -3 \\ -3 & 5 \end{pmatrix}$	$\begin{pmatrix} -3 & 5 \\ 5 & -8 \end{pmatrix}$

2.7.3. The "inverse" Fibonacci Q-matrices

Table 2.4 gives the "direct" and "inverse" Q-matrices. Comparing the "direct" Fibonacci Q-matrices Q^n with the "inverse" Fibonacci Q-matrices Q^{-n}, it is easy to find a very simple method to get the "inverse" matrix Q^{-n} from its "direct" matrix Q^n.

In fact, if the extent n of the "direct" matrix Q^n, given by (2.83), is even ($n=2k$) then for obtaining of the inverse matrix Q^{-n} it is necessary to interchange the places of the diagonal elements F_{n+1} и F_{n-1} in (2.83) and to take the diagonal elements F_n in (2.83) with the opposite sign. This means that for the case $n=2k$ the "inverse" matrix Q^{-n} has the following form:

$$Q^{-n} = \begin{pmatrix} F_{n-1} & -F_n \\ -F_n & F_{n+1} \end{pmatrix} \tag{2.88}$$

To obtain the "inverse" matrix Q^{-n} from the "direct" matrix Q^n, given by (2.83), for the case $n=2k+1$ it is necessary to interchange the places of the diagonal elements F_{n+1} и F_{n-1} in (2.83) and to take them with the opposite sign, that is:

$$Q^{-n} = \begin{pmatrix} -F_{n-1} & F_n \\ F_n & -F_{n+1} \end{pmatrix}. \tag{2.89}$$

2.8. "The theory of Fibonacci numbers" in modern mathematics

2.8.1. Fibonacci Association

Studies of the French mathematicians Lucas and Binet became a launching pad for the deployment of the researches in this area in modern mathematics. In 1963, a group of the U.S. enthusiasts had created Fibonacci Association. In the same year the Fibonacci Association began publishing the mathematical journal *"The Fibonacci Quarterly,"* and starting since 1984 the Fibonacci Association began to hold regularly (1 time for 2 years) the International Conferences "Fibonacci numbers and their applications." The Fibonacci Association has played a huge role in the development of this research direction in the world and had stimulated the Fibonacci studies in many countries.

The American mathematician **Verner Emil Hoggatt** (1921-1981), professor at San Jose State University (USA) was one of the founders of the Fibonacci Association and the magazine *"The Fibonacci Quarterly."*

Figure 2.6. Verner Emil Hoggatt (1921-1981)

In 1969, the publishing house "Houghton Mifflin" published Verner Hoggatt's book "Fibonacci and Lucas Numbers" [39], which is considered until now one of the best books in the field. Verner Hoggatt made great contribution to the promotion of the researches in the field of Fibonacci numbers. His long and, without doubts, outstanding work as a professor of San Jose State University is celebrated by all his followers. He was a scientific supervisor of many master's theses and wrote many articles on the problem of Fibonacci numbers.

The learned monk Brother **Alfred Brousseau** (1907-1988) was another prominent person, involved in the creation of the Fibonacci Association and *The Fibonacci Quarterly.*

Figure 2.7. Alfred Brousseau (1907-1988)

In 1937 he got PhD's degree at the University of California.

At the study of the history of the Fibonacci Association, which has put a rather strange purpose to study the numerical sequence, introduced by the Italian mathematician Fibonacci in the 13th century, there arise the following questions:

1. What is the cause of the increased interest of the members of the Fibonacci Association and many "mathematics lovers" in Fibonacci numbers?

2. What united together two very different people: the mathematician Verner Hoggatt and the representative of the spiritual brotherhood Alfred Brousseau, when they had decided to create the *Fibonacci Association* and to establish the mathematical journal with the unusual title *"The Fibonacci Quarterly"*?

Unfortunately, the direct answer to this question in the short biographies and works of Verner Hoggatt and Alfred Brussau, is absent. But we can try to answer these questions indirectly, by analyzing some of the documents, in particular, their photographs, as well as their books and articles, published in *"The Fibonacci Quarterly."*

In 1969, the magazine "TIME" published the article "The Fibonacci Numbers," dedicated to the Fibonacci Association. This article presented Alfred Brousseau's photo with a cactus in his hands. As is known, the cactus is one of the most characteristic "Fibonacci" botanical objects. The article told about other natural forms, which use Fibonacci numbers, for example, that the Fibonacci numbers are found in spiral formations of many sunflowers, pine cones, branching patterns of trees, and in the arrangement of leaves on the branches of trees and so on.

Alfred Brousseau had recommended to the lovers of Fibonacci numbers *"pay attention to the search for aesthetic satisfaction in them. There is some kind of mystical connection between these numbers and the Universe."*

But on the above Verner Hoggatt's photo (Fig.2.6), he keeps pine cone in his hands. As is known, the pine cone is one of Fibonacci's botanical objects. Hence, from this comparison we can make the assumption that Verner Hoggatt, like Alfred Brousseau, believed in the mystical connection between Fibonacci numbers and the Universe. This belief united the mathematician Verner Hoggatt and learned monk Alfred Brousseau and became their main motive for the deployment of the works on Fibonacci numbers and their applications in modern science.

But, as stated above, Fibonacci numbers are associated with the "golden ratio," because the ratio of the two adjacent Fibonacci numbers in the limit strives for the "golden ratio." This means that the Fibonacci numbers, as the "golden ratio," are expressing the Universe Harmony, that is, indeed, *"there is some kind of mystical connection between these numbers and the Universe"* (Alfred Brousseau).

This means that the theory of Fibonacci numbers, which began to develop rapidly especially since the creation of the Fibonacci Association (1963), **was aimed primarily at solving problems of the harmonization of theoretical natural sciences (physics, chemistry, botany, biology, physiology, medicine and so on), as well as economy, computer science, education and fine arts, related to the golden ratio and Fibonacci numbers.**

This means that the "Harmony of Nature " underlies the theory of Fibonacci numbers [38-40]. The "Problem of Harmony" brings together the above diverse fields of science, economics, computer science, the fine arts and education, which all have relations to Fibonacci numbers and "golden ratio."

Thus, analyzing the causes of the origin of the Fibonacci numbers theory in modern mathematics, we suddenly came to the ancient Greek **"Doctrine on the Numerical Harmony of the Universe,"** which in modern mathematics is embodied in the "theory of Fibonacci numbers"! In this is the essence of the harmonization of mathematics and theoretical natural sciences. And perhaps, it would be right and fair to name this mathematical theory as the **"Mathematical Theory of the Universe Harmony,"** and not to hide the main goal of this theory under strange title of the "Fibonacci numbers theory" [38-40].

2.8.2. A role of Nikolay Vorobyov in the development of the Fibonacci numbers theory

The creation of the American Fibonacci Association and *"The Fibonacci Quarterly"* (1963) is undisputable merit of Professor **Verner Hoggatt** and his associates. However, in fairness, it should be noted that Soviet mathematician **Nikolay Vorobyov** (1925-1995) was the first modern mathematician, who drew attention to the "Fibonacci numbers theory."

Figure 2.8. Nikolay Vorobyov (1925-1995)

In 1961, he published the book "Fibonacci numbers" [38], which had played certainly a very prominent role in the development of the "Fibonacci numbers theory." This book became a best-seller of the 20th century and had a big number of editions, the book was translated into many languages and became a handbook of many Soviet and foreign scientists.

2.8.3. Slavonic "Golden" Group and International Club of the Golden Section

In 1992, in Kiev (Ukraine), the 1st International Workshop **"The Golden Section and Problems of System Harmony"** was hold. The Ukrainian mathematician **Alexey Stakhov**, the Belarusian philosopher **Eduard Soroko** (Minsk), the Ukrainian architect **Oleg Bodnar** (Lvov), the Ukrainian economist **Ivan Tkachenko**, the Russian mechanic **Victor Korobko** (Stavropol), the

Ukrainian chemist **Nikolai Vasyutinskii** (Zaporozhye), the Polish scholar and journalist **Jan Grzhedzelsky** became active participants of the Workshop and the members of the organizing committee. This group of scientists formed the backbone of the informal association of Slavonic scholars, known in the history of science as the "Slavonic "Golden" Group."

In 2003 this group was transformed into the **International Club of the Golden Section** http://www.goldensectionclub.net/ . At the initiative of the Club, the **Institute of the Golden Section**, Academy of Trinitarism (Russia) has been organized in 2005. At the initiative of the Club, the **International Congress on the Mathematics of Harmony** has been held in Odessa (Ukraine) in 2010.

2.8.4. Hilbert's Tenth Problem

As is know, Hilbert's Tenth Problem [27] is called *"The problem of the solution of Diophantine equations."* It dates back in its origin to the ancient mathematician Diophantus.

Hilbert's Tenth Problem was solved by a young Russian mathematician **Yuri Matiyasevich**. His name became widely known in 1970, when he obtained the "negative solution" of Hilbert's Tenth Problem.

The use of the "Fibonacci numbers theory" as presented by Nikolay Vorobyov is the most surprising in Matijasevic's study.

Assessing the impact of research results of Nikolay Vorobyov and American mathematician Julia Robinson on the solution of Hilbert's Tenth Problem, Matiyasevich wrote:

"My original proof of Hilbert's Tenth Problem was based on the theorem, proved in 1942 by the Soviet mathematician Nikolay Vorobyov. This theorem was published only in the 3d expanded edition of Vorobyov's popular book. .... When I have read Julia Robinson's article, I immediately saw that Vorobyov's theorem can be very useful. Julia Robinson did not see the 3d edition of Vorobyov's book as long as she had received a copy of this book from me in 1970. Who could tell what would have happened if Vorobyov had included his theorem into the first edition of his book (1961)? It is possible that Hilbert's Tenth Problem had been solved a decade earlier!"

In the development of Matijasevic's consideration, we can ask the following question: what would have happened if the Italian mathematician Fibonacci did not introduce the Fibonacci numbers in the 13th century? Possibly,

Hilbert's Tenth Problem was not solved until now. Of course, Vorobyov's theorem, used by Matiyasevich, is an important mathematical result, but we should recognize the Italian mathematician Leonardo of Pisa (by the nicknamed Fibonacci) as the main "culprit" of Hilbert's Tenth Problem's solution. In 1202 Fibonacci had published the book "Liber abaci," where he had introduced the new numerical sequence, called Fibonacci numbers.

The main conclusion from these arguments is that the solution of one of the most complicated mathematical problems – Hilbert's Tenth Problem, was obtained by using the Fibonacci numbers theory [38-40]. And this fact is raising on a high level as the Fibonacci numbers theory [38-40] and as the "Mathematics of Harmony" [19].

Chapter 3

HYPERBOLIC FIBONACCI AND LUCAS FUNCTIONS , AND "BODNAR'S GEOMETRY"

3.1. A definition and a brief history of the classical hyperbolic functions

3.1.1. Basic formulas

The function

$$sh(x) = \frac{e^x - e^{-x}}{2} \tag{3.1}$$

is called *hyperbolic sine* and the function

$$ch(x) = \frac{e^x + e^{-x}}{2} \tag{3.2}$$

is called *hyperbolic cosine*.

There is a similarity between trigonometric and hyperbolic functions. By analogy with the trigonometric functions we can define *hyperbolic tangent* and *cotangent*:

$$th(x) = \frac{sh(x)}{ch(x)}, \quad cth(x) = \frac{ch(x)}{sh(x)}.$$

The analytical definitions (3.1), (3.2) can be used to obtain some important identities of the hyperbolic trigonometry. As is well known, there are trigonometric identities, for example, the Pythagorean Theorem for the trigonometric functions:

$$\cos^2\alpha + \sin^2\alpha = 1. \tag{3.3}$$

We can prove similar identities for the hyperbolic functions:

$$ch^2 x - sh^2 x = \left(\frac{e^x + e^{-x}}{2}\right)^2 - \left(\frac{e^x - e^{-x}}{2}\right)^2 = \frac{e^{2x} + 2 + e^{-2x}}{4} - \frac{e^{2x} - 2 + e^{-2x}}{4} = 1$$

Another important property of the hyperbolic functions (3.1), (3.2) is a "parity property":

$$sh(-x) = -sh(x); \quad ch(-x) = ch(x); \quad th(-x) = th(x). \tag{3.4}$$

This means that the hyperbolic sine (3.1) and hyperbolic tangent are odd functions and the hyperbolic cosine (3.2) is an even function.

3.1.2. A history of hyperbolic functions

Although **Johann Heinrich Lambert** (1728-1777), the French mathematician, is often credited for the introduction of the hyperbolic functions, it was actually **Vincenzo Riccati** (1707-1775), the Italian mathematician, who did this in the middle of the 18th century. He studied these functions and used them to obtain solutions of *cubic equations*. Riccati found the standard *addition* formulas, similar to trigonometric identities, for the hyperbolic functions as well as their derivatives. He revealed the relationship between the hyperbolic functions and the exponential function. Riccati first has used the designations *sh* and *ch* for hyperbolic sine and cosine.

The interest in the hyperbolic functions (3.1), (3.2) significantly increased in 19th century, when the Russian geometer **Nikolay Lobachevski** (1792 - 1856) used them to describe mathematical relationships for the non-Euclidian geometry. Because of this, Lobachevski's geometry is also called *hyperbolic geometry*.

3.2. A geometric theory of the hyperbolic functions

3.2.1. Compression and expansion

We start to study a geometric theory of hyperbolic functions from some important geometric transformations [49]. We consider at first the geometric transformation named *compression to a straight line* with the *compression coefficient k* (Fig.3.1-a). This transformation consists in the following. Every point A of a plane passes on into the point A', which lies on the ray PA perpendicular to the straight line *o*, here, the ratio $PA' : PA = k$ or $PA' = k \times PA$ (Fig.3.1-*a*). If the compression coefficient $k > 1$, then $PA' > PA$ (Fig.3.1-*b*); in this case the transformation could be named an *expansion from a straight line*. It is clear that the expansion with the coefficient k is equivalent to the *compression* with the coefficient $\frac{1}{k}$.

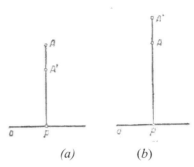

(a)   *(b)*

Figure 3.1. A compression of the point A to a straight line (a) and an expansion from the straight line (b)

A *compression* and an *expansion* possess a number of the important properties [49]:

1. At the compression (expansion) every straight line passes on into straight line
2. At the compression (expansion) all parallels pass on into parallels.
3. At the compression (expansion) the ratio of the segments, lying on one straight line, remains constant
4. At the compression (expansion) the areas of all figures change in the constant ratio equal to the compression coefficient k.

3.2.2. Hyperbola

Let $a > 0$ be some real number. Consider an important geometric curve, called a *hyperbola*. It is described by the following equality:

$$y = \frac{a}{x} \quad \text{or} \quad xy = a \tag{3.5}$$

A graph of the hyperbola is represented in Fig.3.2. It follows from (3.5) and Fig.3.2 that the graph of the hyperbola consists of two branches, which are located for the case $a>0$ in the first quadrant (x and y are positive) and in the third quadrant (x and y are negative) of the coordinate system. Geometrically, the branches of the hyperbola strive for the coordinate axis's, but they are never intersect with them. This means that the coordinate axis's are *asymptotes* of the hyperbola. Note that the equation $xy=a$ is of a simple geometrical interpretation: *the area of the rectangle MQOP, which are bounded by the coordinate axis's and the straight lines, drawn through point M of the hyperbola in parallel to the coordinate axis's (Fig. 3.2), is equal to a*, that is, this area does not depend on the choice of the point M. We will name rectangle $MQOP$ with the area, equal to a, the *coordinate rectangle of the point M*. Then we can give the following geometric definition of the hyperbola:

"A hyperbola is a geometric place of the points, lying in the first and third quadrants of the coordinate system, with the coordinate rectangles, which have constant area".

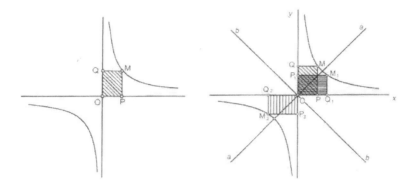

Figure 3.2. Hyperbola   Figure 3.3. The axis's of the hyperbola

It is easy to see that the origin O of the coordinates is a *centre of symmetry* of the hyperbola, that is, the branches of the hyperbola are symmetric one to another, relative to the origin O of the coordinates. The hyperbola has also the *axis's of symmetry*, the bisectors aa and bb of the coordinate angles (Fig.3.3). The centre of symmetry O and the axis's of symmetry aa and bb frequently are called simply *center* and *axis's* of hyperbola. The points of intersection of the hyperbola branches with the axis aa, are called *tops* of the hyperbola.

Hereinafter, we will use analogies between hyperbola and circle. With this purpose, first of all, we introduce a concept of the *diameter* of the hyperbola; it is the line, which passes through the centre of the hyperbola and connects the points of the opposite branches of the hyperbola (it is similar to the diameter of a circle passing through its centre). Let us introduce also a concept of the *radius* of the hyperbola; it is the line, going from the centre of the hyperbola up to the crossing point with the hyperbola (i.e. the radiuses of the hyperbola are determined similarly to the radiuses of a circle).

3.2.3. Geometric definition of the hyperbolic functions

The hyperbola in Fig.3.2 is a base for the geometric definition of the hyperbolic functions. A geometric theory of the *hyperbolic functions* is similar to a theory of traditional (circular) trigonometric functions. In order to emphasize an analogy between circular and hyperbolic trigonometric functions, we will present the theory of hyperbolic functions in parallel with the theory of circular trigonometric functions. We can choose the axis's of symmetry of the hyperbola by the coordinate axis's, as is shown in Fig.3.4, and then use such geometric representation of the hyperbola for the geometric definition of the hyperbolic functions.

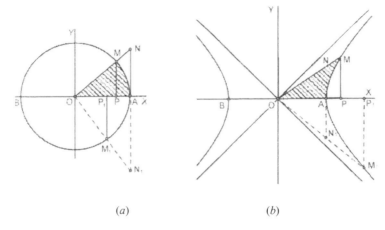

(a)   (b)

Figure 3.4. The unit circle (a) and the unit hyperbola (b)

Let us examine the unit circle (Fig.3.4-*a*)
$$X^2 + Y^2 = 1$$
We can see in Fig.3.4-a the *circular sector OMA*, bounded by the radiuses *OM, OA* and the arc *MA*. The number, equal to the length of the arc *AM* or equal to the double area of the sector bounded by the radiuses *OM* and *OA* and the arc

Let us examine the unit hyperbola (Fig.3.4-*b*)
$$X^2 - Y^2 = 1$$
We can see in Fig.3.4-b the *hyperbolic sector OMA* bounded by the hyperbolic radiuses *OM, OA* and the hyperbolic arc *MA*. Then, the number, equal to the double area of the hyperbolic sector, bounded by these radiuses *OM, OA* and the arc *MA* of the hyperbola, is

MA, is called an *radian angle* α between the radiuses *OA* and *OM* of the circle.

We can drop now the perpendicular *MP* on the diameter *OA* from the point *M* of the circle; in the point *A* we draw a tangent to the circle up to its intersection with the diameter *OM* in the point *N*. The line segment *PM* of the perpendicular is called a *line of sine*, a line segment *OP* of the diameter is called a *line of cosine* and a line segment *AN* is called a *line of tangent*.

called a *hyperbolic angle t* between the hyperbolic radiuses *OA* and *OM*.

We can drop now the perpendicular *MP* from the point *M* of the hyperbola on the diameter *OA*, which is a symmetry axis, intersecting the hyperbola in the top *A*. Then draw a tangent to the hyperbola to its intersection with the radius *OM* in the point *N*. A line segment *PM* of the perpendicular is called a *line of hyperbolic sine*, a line segment *OP* of the axis *X* is called a *line of hyperbolic cosine* and a line segment *AN* is called a *line of hyperbolic tangent*.

The lengths of the line segments *PM, OP* and *AN* are equal respectively to *sine*, *cosine* and *tangent* of the angle α, that is,

$$PM = \sin \alpha, \ OP = \cos \alpha, \ AN = \operatorname{tg} \alpha.$$

The lengths of the line segments *PM, OP* and *AN* are equal respectively to *hyperbolic sine*, *hyperbolic cosine* and *hyperbolic tangent* of the hyperbolic angle *t*, that is,

$$PM = \operatorname{sh} t, \ OP = \operatorname{ch} t, \ AN = \operatorname{th} t.$$

It is known, that the circular trigonometric functions change periodically with the period of 2π. In contrast to this the hyperbolic functions are *not periodic*. It follows from Fig.3.4-*b* that the hyperbolic angle *t* changes from 0 up to ∞. It follows from the definition of the hyperbolic functions what at the change of the hyperbolic angle from 0 up to ∞, the hyperbolic sine sh *t* is changing from 0 up to ∞, the hyperbolic cosine ch *t* is changing from 1 up to ∞ and the hyperbolic tangent th *t* is changing from 0 up to 1. The graphs of these functions are represented in Fig.3.5.

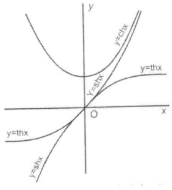

Figure 3.5. Graphs of the hyperbolic functions

By basing on the geometric approach, it is easy to obtain the basic relations for the circular and hyperbolic trigonometric functions.

It follows from a similarity of the triangles OMP and ONA (Fig.3.4-a) that

$$\frac{AN}{OA} = \frac{PM}{OP}.$$

But $\dfrac{AN}{OA} = \text{tg}\alpha$ (because $OA=1$), and

$$\frac{PM}{OP} = \frac{\sin\alpha}{\cos\alpha}.$$

Thus, we have: $\text{tg } \alpha = \dfrac{\sin\alpha}{\cos\alpha}$.

Further, the coordinates of any point M of the circle are equal $OP=X$, $PM=Y$. But then it follows from the unit circle equation $X^2 + Y^2 = 1$ the following important identity:

$$OP^2 + PM^2 = 1$$

or

$$\cos^2\alpha + \sin^2\alpha = 1.$$

By dividing both parts of the obtained identity at first on $\cos^2\alpha$, and then on $\sin^2\alpha$, we get the following remarkable formulas for the trigonometric functions:

$$1 + tg^2\alpha = \frac{1}{\cos^2\alpha};$$

$$ctg^2\alpha + 1 = \frac{1}{\sin^2\alpha}.$$

It follows from a similarity of the triangles OMP and ONA (Fig.3.4-b) that

$$\frac{AN}{OA} = \frac{PM}{OP}.$$

But $\dfrac{AN}{OA} = \text{tg}t$ (because $OA=1$), and

$$\frac{PM}{OP} = \frac{sht}{cht}.$$

Thus, we have: $\text{tg } t = \dfrac{sht}{cht}$.

Further, the coordinates of any point M of the hyperbola are equal $OP=X$, $PM=Y$. But then it follows from the unit hyperbola equation $X^2 - Y^2 = 1$ the following important identity:

$$OP^2 - PM^2 = 1$$

or

$$ch^2 t - sh^2 t = 1.$$

By dividing both parts of the obtained identity at first on ch^2t, and then on sh^2t, we get the following remarkable formulas for the hyperbolic functions:

$$1 - th^2t = \frac{1}{ch^2t};$$

$$ch^2t - 1 = \frac{1}{sh^2t}.$$

By using the geometric approach, we can prove other identities for the hyperbolic functions, in particular:

$$\text{sh}(t+u) = \text{sh}t\ \text{ch}u + \text{ch}t\ \text{sh}u;$$

$$\text{ch}(t+u) = \text{ch}t\ \text{ch}u + \text{sh}t\ \text{sh}u;$$

$$\text{sh } 2t = 2\ \text{sh}t\ \text{ch}t;$$

$$\text{ch } 2t = \text{ch}^2t + \text{sh}^2t;$$

$$\text{th } 2t = \frac{2tht}{1 + th^2t}.$$

3.2.4. Hyperbolic rotation

We continue to study the hyperbola $xy=a$. First of all, we make the compression of a plane to the axis x with the compression coefficient k. In this case the hyperbola $xy=a$ passes on into the hyperbola $xy=ak$, because the abscissa x remains without change and the ordinate y is replaced by yk. Then, we make one more compression of a plane already to the axis y with the coefficient $\frac{1}{k}$.

Note that the compression with the coefficient $\frac{1}{k}$ is equivalent to the expansion with the coefficient k. After the fulfilment of the compression to the axis y with the coefficient $\frac{1}{k}$, what is equivalent to the extension from the axis y with the same coefficient k, the hyperbola $xy=ak$ passes on into the hyperbola $xy = \frac{ak}{k} = a$, because the ordinate y of each point for the case of the new compression to the axis y does not vary, and the abscissa x passes on into $\frac{1}{k}$. Thus, we can see that the *sequential compression of the plane to the axis x with the compression coefficient k and then to the axis y with the compression coefficient $\frac{1}{k}$ transforms the hyperbola xy=a into itself.* A sequence of these two compressions of the plane to a straight line makes an important geometric transformation called *hyperbolic rotation*. A title of the "hyperbolic rotation" reflects that fact what at such transformation all points of the hyperbola as though "glide on a curve", that is, the hyperbola as though "rotates."

Note once again that the hyperbolic rotation is the *sequential fulfilment of the two geometric transformation, at first the compression of a plane with the coefficient k to the axis x and then the expansion of a plane from the axis y with the same coefficient k.*

It follows from the above properties of the *compression* and *expansion* the following properties of the *hyperbolic rotation*:

1. At the hyperbolic rotation every straight line passes on into a straight line.

2. At the hyperbolic rotation the coordinate axis's (the asymptotes of the hyperbola) pass on into themselves.

3. At the hyperbolic rotation parallels pass on into parallels.

4. At the hyperbolic rotation the ratios of all segments, lying on one and the same straight line, remind constant.

5. At the hyperbolic rotation, the areas of all transferred figures are remind constant.

It is very important to emphasize that *by means of the choice of the appropriate value of the coefficient k, by means of the hyperbolic rotation we can transfer each point of the hyperbola into another point of the same hyperbola.* In fact, the compression to the axis x with the given coefficient k transfers the point (x, y) of the hyperbola $xy=a$ into the point (x, ky) of another hyperbola $xy = ak$; after that, the extension of the point (x, ky) from the axis y with the same coefficient k transfers the point (x, ky) of the hyperbola $xy = ak$ into the point $(\frac{x}{k}, ky)$ of the initial hyperbola.

Thus, as an outcome of the hyperbolic rotation the point (x, y) passes on into the point $(\frac{x}{k}, ky)$ of the initial hyperbola. It follows from here that by means of the suitable hyperbolic rotation we can transfer the point (x, y) of the hyperbola into the point (x_1, y_1) of the same hyperbola, if we take the compression coefficient k as follows:

$$x_1 = \frac{x}{k} \text{ or } k = \frac{x}{x_1}.$$

3.3. Hyperbolic Fibonacci and Lucas functions

3.3.1. A brief history

A history of the hyperbolic Fibonacci and Lucas functions begins from the comparison of the classical hyperbolic functions (3.1), (3.2) with Binet's formulas, deduced by Binet in 19th century. In the book [50] these formulas are represented in the following forms, which are rarely used in mathematics:

$$F_n = \begin{cases} \dfrac{\Phi^n + \Phi^{-n}}{\sqrt{5}} & \text{for } n = 2k+1 \\ \dfrac{\Phi^n - \Phi^{-n}}{\sqrt{5}} & \text{for } n = 2k \end{cases} \tag{3.6}$$

$$L_n = \begin{cases} \Phi^n + \Phi^{-n} & \text{for } n = 2k \\ \Phi^n - \Phi^{-n} & \text{for } n = 2k+1 \end{cases} \tag{3.7}$$

If we compare the hyperbolic functions (3.1), (3.2) with Binet's formulas (3.6), (3.7), we can see that these formulas are similar by their mathematical structure. For the first time, the Fibonacci and Lucas functions have been introduced by the Ukrainian mathematicians **Alexey Stakhov** and **Ivan Tkachenko**. According to the recommendation of academician Yuri Mitropolsky, Stakhov and Tkachenko's article "Hyperbolic Fibonacci trigonometry" [9] has been published in the academic Journal "Reports of the Ukrainian Academy of Sciences" (1993). In the history of mathematics, this article is the first publication, devoted to the hyperbolic Fibonacci and Lucas functions. The functions, introduced in [9], look as follows:

Hyperbolic Fibonacci sine

$$sF(x) = \frac{\Phi^{2x} - \Phi^{-2x}}{\sqrt{5}} \tag{3.8}$$

Hyperbolic Fibonacci cosine

$$cF(x) = \frac{\Phi^{2x+1} + \Phi^{-(2x+1)}}{\sqrt{5}} \tag{3.9}$$

Hyperbolic Lucas sine

$$sL(x) = \Phi^{2k+1} - \Phi^{-(2k+1)} \tag{3.10}$$

Hyperbolic Lucas cosine

$$cL(x) = \Phi^{2k} + \Phi^{-2k} \tag{3.11}$$

The hyperbolic Fibonacci and Lucas functions (3.8) – (3.11), described in [9], have essential shortcoming in comparison to the classical hyperbolic functions (3.1), (3.2). In contrast to the classical hyperbolic functions, the graph of the Fibonacci cosine (3.9) is *asymmetric* in respect to the axis x, while the graph of the Lucas sine (3.10) is *asymmetric* in respect to the origin of coordinates. This means that the "parity property" does not hold for the hyperbolic Fibonacci cosine (3.9) and hyperbolic Lucas sine (3.10). This confines the area of an effective application of a new class of hyperbolic functions given by (3.8)-(3.11).

To eliminate this shortcoming, **Alexey Stakhov** and **Boris Rozin** have introduced in [10-13] the *symmetric hyperbolic Fibonacci and Lucas functions*.

3.3.2. The symmetric hyperbolic Fibonacci and Lucas functions

The symmetric hyperbolic Fibonacci and Lucas sine and cosine, introduced in [10-13], are a very important step in the development of the general theory of hyperbolic functions [13,14]. They look as follows:

<u>Symmetric hyperbolic Fibonacci sine</u>

$$sFs(x) = \frac{\Phi^x - \Phi^{-x}}{\sqrt{5}} \qquad (3.12)$$

<u>Symmetric hyperbolic Fibonacci cosine</u>

$$cFs(x) = \frac{\Phi^x + \Phi^{-x}}{\sqrt{5}} \qquad (3.13)$$

<u>Symmetric hyperbolic Lucas sine</u>

$$sLs(x) = \Phi^x - \Phi^{-x} \qquad (3.14)$$

<u>Symmetric hyperbolic Lucas cosine</u>

$$cLs(x) = \Phi^x + \Phi^{-x} \qquad (3.15)$$

There are the following simple correlations between the hyperbolic Fibonacci functions (3.12), (3.13) and the hyperbolic Lucas functions (3.14), (3.15):

$$sFs(x) = \frac{sLs(x)}{\sqrt{5}}; \qquad cFs(x) = \frac{cLs(x)}{\sqrt{5}}. \qquad (3.16)$$

By analogy with the classical hyperbolic functions (3.1), (3.2) we can introduce the symmetric hyperbolic Fibonacci and Lucas tangents and cotangents.

<u>Symmetric hyperbolic Fibonacci tangent</u>

$$tFs(x) = \frac{sFs(x)}{cFs(x)} = \frac{\Phi^x - \Phi^{-x}}{\Phi^x + \Phi^{-x}} \qquad (3.17)$$

<u>Symmetric hyperbolic Fibonacci cotangent</u>

$$ctFs(x) = \frac{cFs(x)}{sFs(x)} = \frac{\Phi^x + \Phi^{-x}}{\Phi^x - \Phi^{-x}} \qquad (3.18)$$

<u>Symmetric hyperbolic Lucas tangent</u>

$$tLs(x) = \frac{sLs(x)}{cLs(x)} = \frac{\Phi^x - \Phi^{-x}}{\Phi^x + \Phi^{-x}} \qquad (3.19)$$

$$ctLs(x) = \frac{cLs(x)}{sLs(x)} = \frac{\Phi^x + \Phi^{-x}}{\Phi^x - \Phi^{-x}} \tag{3.20}$$

3.3.3. Parity property

The main advantage of the symmetric hyperbolic Fibonacci and Lucas functions (3.12) - (3.15), introduced in [10-13], in comparison to the functions (3.8) – (3.11) [9] is a preservation of the *parity property*. It is proved in [10] the following *parity properties* of the functions (3.12) – (3.15), (3.17) – (3.20):

$$sFs(-x) = -sFs(x); \quad cFs(-x) = cFs(x) \tag{3.21}$$

$$sLs(-x) = -sLs(x); \quad cLs(-x) = cLs(x) \tag{3.22}$$

$$tFs(-x) = -tF(x); \quad tLs(-x) = -tL(x) \tag{3.23}$$

$$ctFs(-x) = -ctF(x); \quad ctLs(-x) = -ctL(x) \tag{3.24}$$

It follows from (3.16) – (3.17) that the symmetric hyperbolic Fibonacci and Lucas sine's (3.12), (3.14) are odd functions, the symmetric hyperbolic Fibonacci and Lucas cosine's (3.13), (3.15) are even functions, the symmetric hyperbolic Fibonacci and Lucas tangents (3.17), (3.19) and the symmetric hyperbolic Fibonacci and Lucas cotangents (3.18), (3.20) are odd functions.

3.3.4. An uniqueness of the hyperbolic Fibonacci and Lucas functions

As is known, Binet's formulas (3.6), (3.7) set forth the so-called *"extended"* Fibonacci and Lucas numbers (see Table 3.1).

Table 3.1. The "extended" Fibonacci and Lucas numbers

n	0	1	2	3	4	5	6	7	8	9	10
F_n	0	1	1	2	3	5	8	13	21	34	55
F_{-n}	0	1	-1	2	-3	5	-8	13	-21	34	-55
L_n	2	1	3	4	7	11	18	29	47	76	123
L_{-n}	2	-1	3	-4	7	-11	18	-29	47	-76	123

Comparing Binet's formulas (3.6), (3.7) with the symmetric hyperbolic Fibonacci and Lucas functions (3.12) - (3.15), it is easy to see that for the discrete values of the variable x (x=0,±1,±2,±3,…) the functions (3.12), (3.13) coincide with the "extended" Fibonacci numbers calculated according to Binet's formula (3.6), that is,

$$F_n = \begin{cases} sFs(n) & \text{for } n = 2k \\ cFs(n) & \text{for } n = 2k+1 \end{cases} \tag{3.25}$$

and the functions (3.14), (3.15) coincide with the "extended" Lucas numbers calculated according to Binet's formula (3.7), that is,

$$L_n = \begin{cases} cLs(n) & \text{for } n = 2k \\ sLs(n) & \text{for } n = 2k+1 \end{cases} \tag{3.26}$$

where k takes the values from the set k=0,±1,±2,±3,…

Note that the classical hyperbolic functions (3.1), (3.2) do not possess the property similar to (3.25), (3.26).

To demonstrate this property clearly, consider the graphs of the symmetric hyperbolic Fibonacci and Lucas functions, represented in Fig. 3.6, 3.7.

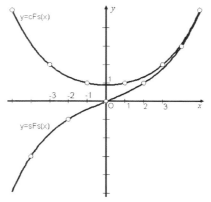

Figure 3.6. Graphs of the symmetric hyperbolic Fibonacci functions

In Fig.3.6, the graphs of the symmetric hyperbolic Fibonacci sine $y = sFs(x)$ and the symmetric hyperbolic Fibonacci cosine $y = cFs(x)$ are represented.

The points on the graph $y = sFs(x)$ correspond to the "extended" Fibonacci numbers with the even indexes $2n$:

$$F_{2n} = \{\ldots, F_{-8} = -21, F_{-6} = -8, F_{-4} = -3, F_{-2} = -1, F_2 = 1, F_4 = 3, F_6 = 8, F_8 = 21, \ldots\}. \tag{3.27}$$

The points on the graph $y = cFs(x)$ correspond to the "extended" Fibonacci numbers with the odd indexes $2n+1$:

$$F_{2n+1} = \{..., F_{-7} = 13, F_{-5} = 5, F_{-3} = 2, F_{-1} = 1, F_1 = 1, F_3 = 3, F_5 = 5, F_7 = 13, ...\} \qquad (3.28)$$

In Fig.3.7, the graphs of the symmetric hyperbolic Lucas sine $y = sLs(x)$ and the symmetric hyperbolic Lucas cosine $y = cLs(x)$ are represented.

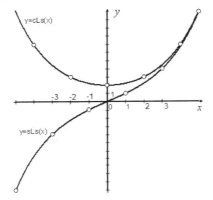

Figure 3.7. The graph of the symmetric hyperbolic Lucas functions

The points on the graph $y = sLs(x)$ correspond to the "extended" Lucas numbers with the odd indexes $2n+1$:

$$L_{2n+1} = \{..., L_{-7} = -29, L_{-5} = -11, L_{-3} = -4, L_{-1} = -1, L_1 = 1, L_3 = 4, L_5 = 11, L_7 = 29, ...\}. \qquad (3.29)$$

The points on the graph $y = cLs(x)$ correspond to the "extended" Lucas numbers with the even indexes $2n$:

$$L_{2n} = \{..., L_{-6} = 18, L_{-4} = 7, L_{-2} = 3, L_0 = 2, L_2 = 3, L_4 = 7, L_6 = 18, ...\} \qquad (3.30)$$

Here it is necessary to point out what in the point $x=0$ the symmetric hyperbolic Fibonacci cosine $cFs(x)$ takes the value $cFs(0) = \dfrac{2}{\sqrt{5}}$ (Fig.3.6), and the symmetric hyperbolic Lucas cosine $cLs(x)$ takes the value $cLs(0)=2$ (Fig.3.7). It is also important to emphasize that the "extended" Fibonacci numbers F_n, with the even indexes ($n = 0, \pm2, \pm4, \pm6, ...$) are "inscribed" into the graph of the symmetric hyperbolic Fibonacci sine $sFs(x)$ in the discrete points ($x = 0, \pm2, \pm4, \pm6, ...$) and the "extended" Fibonacci numbers with the odd indexes ($n = \pm1, \pm3,$

±5, …) are "inscribed" into the symmetric hyperbolic Fibonacci cosine $cFs(x)$ in the discrete points ($x = \pm1, \pm3, \pm5$ …). On the other hand, the "extended" Lucas numbers with the even indexes are "inscribed" into the graph of the symmetric hyperbolic Lucas cosine $cLs(x)$ in the discrete points ($x = 0, \pm2, \pm4, \pm6$ …), and the "extended" Lucas numbers with the odd indexes are "inscribed" into the graph of the symmetric hyperbolic Lucas cosine $sLs(x)$ in the discrete points ($x = \pm1, \pm3, \pm5$ …).

Now we can compare the classical hyperbolic functions (3.1), (3.2) with the symmetric hyperbolic Fibonacci functions (3.12)-(3.15). It follows from this comparison that the symmetric hyperbolic Fibonacci and Lucas functions possess all important properties of the classical hyperbolic functions. We can say what the symmetric hyperbolic Fibonacci and Lucas functions have *hyperbolic properties*.

On the other hand, a comparison of the "extended" Fibonacci and Lucas numbers, given by Binet's formulas (3.6), (3.7) with the symmetric hyperbolic Fibonacci and Lucas functions (3.12) – (3.15) show what according to (3.25), (3.26) these functions are a generalization of the "extended" Fibonacci and Lucas numbers for continuous domain. This means that the symmetric hyperbolic Fibonacci and Lucas functions possess *recursive properties* similar to the properties of the "extended" Fibonacci and Lucas numbers.

3.4. The hyperbolic properties

3.4.1. The basic theorems

It is proved in [10-13] that for each identity for the classical hyperbolic functions there is an analogue in the form of the corresponding identity for the hyperbolic Fibonacci and Lucas functions. The symmetric hyperbolic Fibonacci and Lucas functions (3.12) – (3.15) have properties, similar to the classical hyperbolic functions (3.1), (3.2). Consider some of them in comparison to the certain properties of the classical hyperbolic functions.

Theorem 3.1. The following relations, similar to the relation $\left[ch(x)\right]^2 - \left[sh(x)\right]^2 = 1$, are valid for the symmetric hyperbolic Fibonacci and Lucas functions:

$$\left[cFs(x)\right]^2 - \left[sFs(x)\right]^2 = \frac{4}{5}. \tag{3.31}$$

$$\left[cLs(x)\right]^2 - \left[sLs(x)\right]^2 = 4. \tag{3.32}$$

Proof:

The relation (3.31)

$$\left[cFs(x)\right]^2 - \left[sFs(x)\right]^2 = \left(\frac{\Phi^x + \Phi^{-x}}{\sqrt{5}}\right)^2 - \left(\frac{\Phi^x - \Phi^{-x}}{\sqrt{5}}\right)^2$$
$$= \frac{\Phi^{2x} + 2 + \Phi^{-2x} - \Phi^{2x} + 2 - \Phi^{-2x}}{5} = \frac{4}{5} \tag{3.33}$$

The relation (3.32) is proved in a similar manner as (3.33).

Theorem 3.2. The following relations, similar to the relation $ch(x+y) = ch(x)ch(y) + sh(x)sh(y)$, are valid for the symmetric hyperbolic Fibonacci and Lucas functions:

$$\frac{2}{\sqrt{5}} cFs(x+y) = cFs(x)cFs(y) + sFs(x)sFs(y) \tag{3.34}$$

$$2cLs(x+y) = cLs(x)cLs(y) + sLs(x)sLs(y) \tag{3.35}$$

Proof:

The relation (3.34)

$$cFs(x)cFs(y) \pm sFs(x)sFs(y) =$$
$$= \frac{\Phi^x + \Phi^{-x}}{\sqrt{5}} \times \frac{\Phi^y + \Phi^{-y}}{\sqrt{5}} + \frac{\Phi^x - \Phi^{-x}}{\sqrt{5}} \times \frac{\Phi^y - \Phi^{-y}}{\sqrt{5}} =$$
$$= \frac{\left(\Phi^{x+y} + \Phi^{x-y} + \Phi^{-x+y} + \Phi^{-x-y}\right) + \left(\Phi^{x+y} - \Phi^{x-y} - \Phi^{-x+y} + \Phi^{-x-y}\right)}{5} = \tag{3.36}$$
$$= \frac{2\left(\Phi^{x+y} + \Phi^{-x-y}\right)}{\sqrt{5} \times \sqrt{5}} = \frac{2}{\sqrt{5}} cFs(x+y)$$

The relation (3.35) is proved in a similar manner as (3.36).

Theorem 3.3. The following relations, similar to the relation $ch(x-y) = ch(x)ch(y) - sh(x)sh(y)$, are valid for the symmetric hyperbolic Fibonacci and Lucas functions:

$$\frac{2}{\sqrt{5}} cFs(x-y) = cFs(x)cFs(y) - sFs(x)sFs(y) \tag{3.37}$$

$$2cLs(x-y) = cLs(x)cLs(y) - sLs(x)sLs(y). \tag{3.38}$$

The correlations (3.37), (3.38) are proved in a similar manner as (3.36).

In the similar manner, we can prove the following theorems for the symmetric hyperbolic Fibonacci and Lucas functions.

Theorem 3.4. The following relations, similar to the relation $ch(2x) = \left[ch(x)\right]^2 + \left[sh(x)\right]^2$, are valid for the symmetric hyperbolic Fibonacci and Lucas functions:

$$\frac{2}{\sqrt{5}} cFs(2x) = \left[cFs(x)\right]^2 + \left[sFs(x)\right]^2 \tag{3.38}$$

$$2cLs(2x) = \left[cLs(x)\right]^2 + \left[sLs(x)\right]^2. \tag{3.39}$$

Theorem 3.5. The following relations, similar to the relation $sh(2x) = 2sh(x)ch(x)$, are valid for the symmetric hyperbolic Fibonacci and Lucas functions:

$$\frac{1}{\sqrt{5}} sFs(2x) = sFs(x)cFs(x) \tag{3.40}$$

$$sLs(2x) = sLs(x)cLs(x) \tag{3.41}$$

Theorem 3.6. The following formula, similar to Moivre's formulas $\left[ch(x) \pm sh(x)\right]^n = ch(nx) \pm sh(nx)$, is valid for the symmetric hyperbolic Fibonacci functions:

$$\left[cFs(x) \pm sFs(x)\right]^n = \left(\frac{2}{\sqrt{5}}\right)^{n-1} [cF(nx) \pm sFs(nx)]. \tag{3.42}$$

In Table 3.2 some formulas for the classical hyperbolic functions and the corresponding formulas for the hyperbolic Fibonacci functions are represented.

Table 3.2. "Hyperbolic" properties for the hyperbolic Fibonacci functions

Formulas for the classical hyperbolic functions	Formulas for the hyperbolic Fibonacci functions
$sh(x)=\dfrac{e^{x}-e^{-x}}{2}; ch(x)=\dfrac{e^{x}+e^{-x}}{2}$	$sFs(x)=\dfrac{\Phi^{x}-\Phi^{-x}}{\sqrt{5}}; cFs(x)=\dfrac{\Phi^{x}+\Phi^{-x}}{\sqrt{5}}$
$ch^{2}(x)-sh^{2}(x)=1$	$\left[cFs(x)\right]^{2}-\left[sFs(x)\right]^{2}=\dfrac{4}{5}$
$sh(x+y)=sh(x)ch(x)+ch(x)sh(x)$ $sh(x-y)=sh(x)ch(x)-ch(x)sh(x)$	$\dfrac{2}{\sqrt{5}}sFs(x+y)=sFs(x)cFs(x)+cFs(x)sFs(x)$ $\dfrac{2}{\sqrt{5}}sFs(x-y)=sFs(x)cFs(x)-cFs(x)sFs(x)$
$ch(x+y)=ch(x)ch(x)+sh(x)sh(x)$ $ch(x-y)=ch(x)ch(x)-sh(x)sh(x)$	$\dfrac{2}{\sqrt{5}}cFs(x+y)=cFs(x)cFs(x)+sFs(x)sFs(x)$ $\dfrac{2}{\sqrt{5}}cFs(x-y)=cFs(x)cFs(x)-sFs(x)sFs(x)$
$ch(2x)=2sh(x)ch(x)$	$\dfrac{1}{\sqrt{5}}cFs(2x)=sFs(x)cFs(x)$
$\left[ch(x)\pm sh(x)\right]^{n}=ch(nx)\pm sh(nx)$	$\left[cFs(x)\pm sFs(x)\right]^{n}=\left(\dfrac{2}{\sqrt{5}}\right)^{n-1}\left[cFs(nx)\pm sFs(nx)\right]$

3.4.2. Some useful relations

It is easy to prove the following relation:

$$\left(\Phi_{\lambda}\right)^{x}=e^{x\ln\Phi_{\lambda}}. \tag{3.43}$$

By using (3.43), we can prove the following relations between hyperbolic Fibonacci functions and classical hyperbolic functions:

$$\begin{cases} sF_{\lambda}(x)=\dfrac{2}{\sqrt{4+\lambda^{2}}}sh\left[(\ln\Phi_{\lambda})x\right]; \\ cF_{\lambda}(x)=\dfrac{2}{\sqrt{4+\lambda^{2}}}ch\left[(\ln\Phi_{\lambda})x\right] \end{cases} \tag{3.44}$$

$$\begin{cases} sh(x)=\dfrac{\sqrt{4+\lambda^{2}}}{2}sF_{\lambda}\left(\dfrac{x}{\ln\Phi_{\lambda}}\right); \\ ch(x)=\dfrac{\sqrt{4+\lambda^{2}}}{2}cF_{\lambda}\left(\dfrac{x}{\ln\Phi_{\lambda}}\right) \end{cases} \tag{3.45}$$

We can use the formulas (3.44), (3.45) for differentiation and integration and the withdrawal of other relations. Hence, for example, we have immediately:

$$\begin{cases} \dfrac{d\left[sF_\lambda(x)\right]}{dx} = (\ln \Phi_\lambda)cF_\lambda(x); \\[3mm] \dfrac{d\left[cF_\lambda(x)\right]}{dx} = (\ln \Phi_\lambda)sF_\lambda(x) \end{cases} \tag{3.46}$$

$$\begin{cases} \displaystyle\int sF_\lambda(x)\,dx = \dfrac{cF_\lambda(x)}{\ln \Phi_\lambda} + const; \\[3mm] \displaystyle\int cF_\lambda(x)\,dx = \dfrac{sF_\lambda(x)}{\ln \Phi_\lambda} + const \end{cases} \tag{3.47}$$

$$\left[cF_\lambda(x)\right]^2 - \left[sF_\lambda(x)\right]^2 = \frac{4}{4+\lambda^2} \tag{3.48}$$

3.5. The recursive properties

3.5.1. The basic theorems

The recursive properties of the symmetric hyperbolic Fibonacci and Lucas functions are one more confirmation of the uniqueness of this class of hyperbolic functions, because the classical hyperbolic functions (3.1), (3.2) do not have such a property.

Consider now the recursive properties of the symmetric hyperbolic Fibonacci and Lucas functions in comparison to the similar properties of the "extended" Fibonacci and Lucas numbers.

Theorem 3.7. The following relations, which are similar to the recursive relation for the Fibonacci numbers $F_{n+2} = F_{n+1} + F_n$, are valid for the symmetric hyperbolic Fibonacci functions:

$$sFs(x+2) = cFs(x+1) + sFs(x). \tag{3.49}$$

$$cFs(x+2) = sFs(x+1) + cFs(x) \tag{3.50}$$

Proof:

The relation (3.49)

$$cFs(x+1)+sFs(x)=\frac{\Phi^{x+1}+\Phi^{-x-1}}{\sqrt{5}}+\frac{\Phi^{x}-\Phi^{-x}}{\sqrt{5}}=\frac{\Phi^{x}(\Phi+1)-\Phi^{-x}(1-\Phi)}{\sqrt{5}}=\frac{\Phi^{x}\Phi^{2}-\Phi^{-x}\Phi^{-2}}{\sqrt{5}}=$$

$$=\frac{\Phi^{x+2}-\Phi^{-x-2}}{\sqrt{5}}=sFs(x+2)$$

(3.51)

The relation (3.50) is proved in a similar manner as (3.51).

Theorem 3.8. The following relations, which are similar to the recursive relation for the Lucas numbers $L_{n+2}=L_{n+1}+L_{n}$, are valid for the symmetric hyperbolic Lucas functions:

$$sLs(x+2)=cLs(x+1)+sLs(x).$$ (3.52)

$$cLs(x+2)=sLs(x+1)+cLs(x)$$ (3.53)

The relations (3.52), (3.53) are proved in a similar manner as (3.51).

Theorem 3.9. (a generalization of Cassini's formula). The following relations, which are similar to Cassini's formula $F_{n}^{2}-F_{n+1}F_{n-1}=(-1)^{n+1}$, are valid for the symmetric hyperbolic Fibonacci functions:

$$\left[sFs(x)\right]^{2}-cFs(x+1)cFs(x-1)=-1;$$ (3.54)

$$\left[cFs(x)\right]^{2}-sFs(x+1)sFs(x-1)=1.$$ (3.55)

Proof:

The relation (3.54)

$$\left[sFs(x)\right]^{2}-cFs(x+1)cFs(x-1)=\left(\frac{\Phi^{x}-\Phi^{-x}}{\sqrt{5}}\right)^{2}-\frac{\Phi^{x}-\Phi^{-x}}{\sqrt{5}}\times\frac{\Phi^{x+1}-\Phi^{-x-1}}{\sqrt{5}}=$$

$$=\frac{\Phi^{2x}-2+\Phi^{-2x}-\left(\Phi^{2x}+\Phi^{2}+\Phi^{-2}+\Phi^{-2x}\right)}{5}=\frac{-(2+\Phi+1+2-\Phi)}{5}=-1$$

(3.56)

The relation (3.55) is proved in a similar manner as (3.56).

By using the formulas (3.8) – (3.11), we can prove different identities for the symmetric hyperbolic Fibonacci and Lucas functions. A small part of these identities is presented in Table 3.3.

Table 3.3. The recursive properties of the hyperbolic Fibonacci and Lucas functions

The identities for the Fibonacci and Lucas numbers	The identities for the hyperbolic Fibonacci and Lucas functions
$F_{n-2} = F_{n-1} + F_n$	$sFs(x+2) = cFs(x+1) + sFs(x); cFs(x+2) = sFs(x+1) + cFs(x)$
$L_{n-2} = L_{n-1} + L_n$	$sLs(x+2) = cLs(x+1) + sLs(x); cLs(x+2) = sLs(x+1) + cLs(x)$
$F_n = (-1)^{n-1} F_{-n}$	$sFs(x) = -sFs(-x); cFs(x) = cFs(-x)$
$L_n = (-1)^n L_{-n}$	$sLs(x) = -sLs(-x); cLs(x) = cLs(-x)$
$F_{n-3} + F_n = 2F_{n-2}$	$sFs(x+3) + cFs(x) = 2cFs(x+2); cFs(x+3) + sFs(x) = 2sFs(x+2)$
$F_{n-3} - F_n = 2F_{n-1}$	$sFs(x+3) - cFs(x) = 2sFs(x+1); cFs(x+3) - sFs(x) = 2cFs(x+1)$
$F_{n-6} - F_n = 4F_{n-3}$	$sFs(x+6) - cFs(x) = 4cFs(x+3); cFs(x+6) - sFs(x) = 4cFs(x+3)$
$F_n^2 - F_{n-1}F_{n-1} = (-1)^{n-1}$	$[sFs(x)]^2 - cFs(x+1)cFs(x-1) = -1; [cFs(x)]^2 - sFs(x+1)sFs(x-1) = 1$
$F_{2n-1} = F_{n-1}^2 + F_n^2$	$cFs(2x+1) = [cFs(x+1)]^2 + [sFs(x)]^2; sFs(2x+1) = [sFs(x+1)]^2 + [cFs(x)]^2$
$L_n^2 - 2(-1)^n = L_{2n}$	$[sLs(x)]^2 + 2 = cLs(2x); [cLs(x)]^2 - 2 = cLs(2x)$
$L_n + L_{n-3} = 2L_{n-2}$	$sLs(x) + cLs(x+3) = 2sLs(x+2); cLs(x) + sLs(x+3) = 2cLs(x+2)$
$L_{n-1}L_{n-1} - L_n^2 = -5(-1)^n$	$sLs(x+1)sLs(x-1) - [cLs(x)]^2 = -5; cLs(x+1)cLs(x-1) - [sLs(x)]^2 = 5$
$F_{n-3} - 2F_n = L_n$	$sFs(x+3) - 2cFs(x) = sLs(x); cFs(x+3) - 2sFs(x) = cLs(x)$
$L_{n-1} + L_{n-1} = 5F$	$sLs(x-1) + cLs(x+1) = 5sFs(x); cLs(x-1) + sLs(x+1) = 5cFs(x)$
$L_n + 5F_n = 2L_{n-1}$	$sLs(x) + 5cFs(x) = 2cLs(x+1); cLs(x) + 5sFs(x) = 2sLs(x+1)$
$L_{n-1}^2 + L_n^2 = 5F_{2n-1}$	$[sLs(x+1)]^2 + [cLs(x)]^2 = 5cFs(2x+1); [cLs(x+1)]^2 + [sLs(x)]^2 = 5sFs(2x+1)$

3.5.2. Theory of Fibonacci numbers as a "degenerate" case of the theory of the hyperbolic Fibonacci and Lucas functions

It is shown above, the two "continuous" identities for the hyperbolic Fibonacci and Lucas functions always correspond to every "discrete" identity for the Fibonacci and Lucas numbers. Conversely, one may obtain a "discrete" identity for the Fibonacci and Lucas numbers by using two corresponding "continuous" identities for the hyperbolic Fibonacci and Lucas functions. As the "extended" Fibonacci and Lucas numbers, according to (3.25) and (3.26), are "discrete" cases of the hyperbolic Fibonacci and Lucas functions for the "discrete" values of the continuous variable x, this means that with the introduction of the hyperbolic Fibonacci and Lucas functions the classical "theory of Fibonacci numbers" [38-40] as if "degenerates," because this theory is the special ("discrete") case of the more general ("continuous) theory of the hyperbolic Fibonacci and Lucas functions. This conclusion is another unexpected result, which follows from the theory of the hyperbolic Fibonacci and Lucas functions [10-12]. But the new geometric theory of phyllotaxis, created by the Ukrainian researcher Oleg Bodnar [20-23], is the most brilliant confirmation of the uniqueness and fundamental nature of the hyperbolic Fibonacci and Lucas functions.

3.6. Phyllotaxis phenomenon

3.6.1. Dirac's Principle of Mathematical Beauty

A public lecture: "The complexity of finite sequences of zeros and units, and the geometry of finite functional spaces" [51], presented by the eminent Russian mathematician academician **Vladimir Arnold** at the Moscow Mathematical Society on May 13, 2006, contains many interesting ideas, which touch a beauty of mathematics and evaluation of its role in experimental science. Arnold notes:

1. In my opinion, mathematics is simply a part of physics, that is, it is an experimental science, which discovers for mankind the most important and simple laws of nature.

2. We must begin with a beautiful mathematical theory. Dirac states: *"If this theory is really beautiful, then it necessarily will appear as a fine model of important physical phenomena. It is necessary to search for these phenomena to develop applications of the beautiful mathematical theory and to interpret them as predictions of new laws of physics."* Thus, according to Dirac, all new physics, including relativistic and quantum, are developing in this way.

At Moscow University there is a tradition, when the distinguished scientists are requested to write some self-chosen inscription on a blackboard. When Dirac had visited Moscow in 1956, he had written: *"A physical law must possess mathematical beauty."* This inscription is an essence of the famous *Principle of Mathematical Beauty*, developed by Dirac during his scientific life. No other modern physicist has been preoccupied by the concept of beauty more than Dirac.

Thus, according to Dirac, the *Principle of Mathematical Beauty* is the primary criterion for a mathematical theory to be considered as a model of physical phenomena. Of course, there is an element of subjectivity in the definition of the "beauty" of mathematics, but the majority of mathematicians agrees that "beauty" in mathematical objects and theories nevertheless exist.

3.6.2. What is the phyllotaxis?

Among Nature's phenomena, which surround us, perhaps, the *botanical phenomenon of phyllotaxis* is the best known and most common. This phenomenon is inherent to many biological objects. An essence of phyllotaxis consists in a spiral disposition of leaves on plant's stems of trees, petals in flower baskets, seeds in pine cone and sunflower head etc. This phenomenon, known already to Kepler, was a subject of discussion of many scientists, including Leonardo da Vinci, Turing, Veil and so on. In phyllotaxis phenomenon there are used more complex concepts of symmetry, in particular, a concept of *helical symmetry*.

The phenomenon of phyllotaxis reveals itself especially brightly in inflorescences and densely packed botanical structures such, as pine cones, pineapples, cacti, heads of sunflower and cauliflower and many other botanical objects (Fig.3.8).

(a)   (b)   (c)

(d)   (e)   (f)

Figure 3.8. Phyllotaxis structures: (a) cactus; (b) head of sunflower; (c) coneflower; (d) Romanescue cauhflower; (e) pineapple; (f) pinecone

On the surfaces of such objects, their bio-organs (seeds on the disks of sunflower heads and pine cones etc.) are placed in the form of the left-twisted and right-twisted spirals. For such phyllotaxis objects, it is used usually the number ratios of the left-hand and right-hand spirals, observed on the surface of the phyllotaxis objects. Botanists proved that these ratios are equal to the ratios of the adjacent Fibonacci numbers, that is,

$$\frac{F_{n+1}}{F_n}: \quad \frac{2}{1}, \frac{3}{2}, \frac{5}{3}, \frac{8}{5}, \frac{13}{8}, \frac{21}{13}, \dots \rightarrow \Phi = \frac{1+\sqrt{5}}{2}. \tag{3.57}$$

The ratios (3.57) are called *phyllotaxis orders* [8]. They are different for different phyllotaxis objects. For example, a head of sunflower can have the phyllotaxis orders given by Fibonacci's ratios $\frac{89}{55}, \frac{144}{89}$ and even $\frac{233}{144}$.

Geometric models of phyllotaxis structures in Fig.3.9 give more clear representation about this unique botanical phenomenon.

(a)

(b)

(c)

Figure 3.9. Geometric models of phyllotaxis structures:(a) pineapple; (б) pine cone; (в) head of sunflower

3.6.3. Puzzle of phyllotaxis

By observing the object of phyllotaxis in the final form and enjoying the well organized picture on its surface, we always ask the question: how are Fibonacci's spirals are formed on its surface during its growth? It is proved that a majority of bio-forms changes their phyllotaxis orders during their growth. It is known, for example, that sunflower disks located on the different levels of the same stalk have the different phyllotaxis orders; moreover, the more an age of the disk, the more its phyllotaxis order. This means that during the growth of the phyllotaxis object, a natural modification (an increase) of symmetry happens and this modification of symmetry obeys the law:

$$\frac{2}{1} \to \frac{3}{2} \to \frac{5}{3} \to \frac{8}{5} \to \frac{13}{8} \to \frac{21}{13} \to \dots \tag{3.58}$$

The modification of the phyllotaxis orders according to (3.58) is named *dynamic symmetry* [20]. All the above data are the essence of the well known *"puzzle of phyllotaxis"*. Many scientists, who studied this problem, did believe what the phenomenon of the dynamical symmetry (3.58) is of fundamental interdisciplinary importance. In opinion of **Vladimir Vernadsky**, the famous Russian scientist-encyclopaedist, a problem of biological symmetry is the key problem of biology.

Thus, the phenomenon of the dynamic symmetry (3.58) plays a special role in the geometric problem of phyllotaxis. One may assume that the numerical regularity (3.58) reflects some general geometric laws, which hide a secret of the dynamic mechanism of phyllotaxis, and their uncovering would be of great importance for the understanding of the phyllotaxis phenomenon in the whole. A connection of botanical phenomenon of phyllotaxis with the Fibonacci and Lucas numbers, as well as with other recurrent generalized Fibonacci sequences, significantly enhances the role of "the theory of Fibonacci numbers" in modern science. And in this regard, of particular importance are the hyperbolic Fibonacci and Lucas functions that have very interesting applications for simulation of this phenomenon.

A new geometric theory of phyllotaxis, developed recently by Ukrainian researcher Oleg Bodnar [20], is based on the hyperbolic Fibonacci and Lucas functions.

3.7. Bodnar's geometry

3.7.1. Structural-numerical analysis of phyllotaxis lattices

Let's consider the basic ideas and concepts of Bodnar's geometry [20]. We can see in Fig.3.10-*a* a cedar cone as characteristic example of phyllotaxis' object.

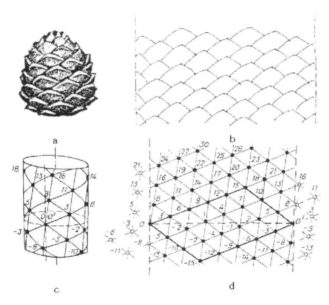

Figure 3.10. Analysis of structure-numerical properties of the phyllotaxis lattice

On the surface of the cedar cone its each seed forms a block with the adjacent seeds in three directions. As the outcome we can see the picture, which consists of three types of spirals; the numbers of the spirals are equal to the Fibonacci numbers: 3, 5, 8. With the purpose of the simplification of the geometric model of the phyllotaxis object in a Fig. 3.10-a, b, we represent the phyllotaxis object in the cylindrical form (Fig.3.10-c). If we cut the surface of the cylinder in Fig.3.10-c by the vertical straight line and then unroll the cylinder on a plane (Fig.3.10-d), we get a fragment of the phyllotaxis lattice, bounded by the two parallel straight lines, which are the traces of the cutting line. We can see that the three groups of parallel straight lines in Fig.3.10-d, namely, the three straight lines 0-21, 1-16, 2-8 with the right-hand small declination; the five straight lines

106

3-8, 1-16, 4-19, 7-27, 0-30 with the left-hand declination; and the eight straight lines 0-24, 3-27, 6-30, 1-25, 4-25, 7-28, 2-18, 5-21 with the right-hand abrupt declination, correspond to three types of spirals on the surface of the cylinder in Fig.3.10-c.

We will use the following method of numbering of the lattice nodes in Fig.3.10-d. Let us introduce now the following system of coordinates. We will use the direct line OO' as the abscissa axis and the vertical trace, which passes through the point O, as the ordinate axis. We will take now the ordinate of the point 1 as the length unit, then the number, ascribed to some point of the lattice, will be equal to its ordinate. The lattice, numbered by the indicated method, has a few characteristic properties. Any pair of the points gives a certain direction in the lattice system and, finally, the set of the three parallel directions of the phyllotaxis lattice. We can see that the lattice in Fig.3.10-d consists of triangles. The vertices of the triangles are numbered by the numbers a, b, c. It is clear that the lattice in Fig.3.10-d consists of the set triangles of the kind $\{c,b,a\}$, for example, $\{0,3,8\}$, $\{3,6,11\}$, $\{3,8,11\}$, $\{6,11,14\}$ an so on. It is important to note that the sides of the triangle $\{c,b,a\}$ are equal to the remainders between the numbers a, b, c of the triangle $\{a, b, c\}$ and are the adjacent Fibonacci numbers: 3, 5, 8. For example, for the triangle $\{0,3,8\}$ we have the following remainders: 3-0=3, 8-3=5, 8-0=8. This means that the sides of the triangle $\{0,3,8\}$ are equal respectively 3, 5, 8. For the triangle $\{3,6,11\}$ we have: 6-3=3, 11-6=5, 11-3=8. This means that its sides are equal 3, 5, 8, respectively. Here each side of the triangle defines one of three declinations of the straight lines, which make the lattice in Fig. 2.3-d. In particular, the side of the length 3 defines the right-hand small declination, the side of the length 5 defines the left-hand declination and the side of the length 8 defines the right-hand abrupt declination. Thus, Fibonacci numbers 3, 5, 8 determines a structure of the phyllotaxis lattice in Fig.3.10-d.

The second property of the lattice in Fig.3.10-d is the following. The line segment OO' can be considered as a diagonal of the parallelogram constructed on the basis of the straight lines corresponding to the left-hand declination and the right-hand small declination. Thus, the given parallelogram allows to evaluate symmetry of the lattice without the use of digital numbering. We will name this parallelogram by *coordinate parallelogram*. Note that the coordinate parallelograms of different sizes correspond to the lattices with different symmetry.

3.7.2. Dynamic symmetry of the phyllotaxis objects

We start the analysis of the phenomenon of dynamic symmetry. The idea of the analysis consists in the comparison of the series of the phyllotaxis lattices (the unrolling of the cylindrical lattice) with different symmetry (Fig. 3.11).

In Fig.3.11 the variant of Fibonacci's phyllotaxis is illustrated, when we observe the following modification of the dynamic symmetry of the phyllotaxis object during its growth:

$$1:2:1 \rightarrow 2:3:1 \rightarrow 2:5:3 \rightarrow 5:8:3 \rightarrow 5:13:8.$$

Note that the lattices, represented in Fig.3.11, are considered as the sequential stages (5 stages) of the transformation of one and the same phyllotaxis object during its grows. There is a question: how are carrying out the transformations of the lattices, that is, which geometric movement can be used to provide the sequential passing all the illustrated stages of the phyllotaxis lattice?

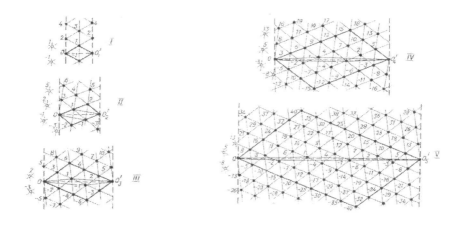

Figure 3.11. Analysis of the dynamic symmetry of phyllotaxis object

3.7.3. The key idea of Bodnar's geometry

We will not go deep into Bodnar's original reasoning's, which resulted him to a new geometrical theory of phyllotaxis, in order to become acquainted with his original geometry more in detailed, we recommend to readers the Bodnar's book [20] and his newest publications on this theme [21-23]. We pay reader's attention only to the two key ideas, which underlie this geometry.

Let us begin from the analysis of the phenomenon of dynamic symmetry. The idea of the analysis consists in the comparison of the series of the phyllotaxis lattices of different symmetry (Fig.3.11). We start from the comparison of the stages I and II. At these stages the lattice can be transformed by the compression of the plane along the direction 0-3 up to the position, when the line segment 0-3 attains the edge of the lattice. Simultaneously the expansion of the plane in the direction 1-2, perpendicular to the compression direction, should happen. At the passing on from the stage II to the stage III, the compression should be made along the direction O-5 and the expansion along the perpendicular direction 2-3. The next passage is accompanied by the similar deformations of the plane in the direction O-8 (compression) and in the perpendicular direction 3-5 (expansion).

But we know that the compression of a plane to any straight line with the coefficient k and the simultaneous expansion of a plane in the perpendicular direction with the same coefficient k are nothing as *hyperbolic rotation* [41]. A scheme of hyperbolic transformation of the lattice fragment is presented in Fig.3.12. The scheme corresponds to the stage II of Fig.3.11. Note that the hyperbola of the first quadrant has the equation $xy=1$, and the hyperbola of the fourth quadrant has the equation $xy=-1$.

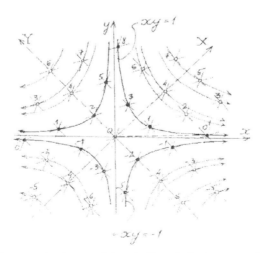

Figure 3.12. A general scheme of the phyllotaxis lattice transformation in the system of the equatorial hyperboles

It follows from this consideration the first key idea of Bodnar's geometry: **the transformation of the phyllotaxis lattice in the process of its growth is carried out by means of the *hyperbolic rotation*, the main geometric transformation of hyperbolic geometry.**

This transformation is accompanied by a modification of dynamic symmetry, which can be simulated by the sequential passage from the object with the smaller symmetry order to the object with the larger symmetry order.

However, this idea does not give the answer to the question: why the phyllotaxis lattices in Fig.3.11 are based on Fibonacci numbers?

3.7.4. The "golden" hyperbolic functions

For more detail study of the metric properties of the lattice in Fig.3.12 we consider its fragment, represented in Fig.3.13. Here the disposition of the points is similar to Fig.3.12.

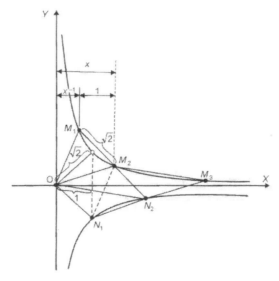

Figure 3.13. The analysis of the metric properties of the phyllotaxis lattice

Let us note the basic peculiarities of the disposition of the points in Fig.3.13:
(1) the points M_1 and M_2 are symmetrical regarding to the bisector of the right angle YOX;

110

(2) the geometric figures $OM_1M_2N_1$, $OM_2N_2N_1$, $OM_2M_3N_2$ are parallelograms;

(3) the point A is the vertex of the hyperbola $yx=1$, that is, $x_A=1$, $y_A=1$, therefore $OA = \sqrt{2}$.

Let us evaluate the abscissa of point M_2 denoted $x_{M_2} = x$. Taking into consideration a symmetry of the points M_1 and M_2, we can write: $x_{M_1} = x^{-1}$. It follows from the symmetry condition of these points what the line segment M_1M_2 is tilted to the coordinate axis's under the angle of 45°. The line segment M_1M_2 is parallel to the line segment ON_1; this means that the line segment ON_1 is tilted to the coordinate axis's under the angle of 45°. Therefore, the point N_1 is a top of the lower branch of the hyperbola; here $x_{N_1} = 1$, $y_{N_1} = 1$, $ON_1 = OA = \sqrt{2}$. It is clear that $ON_1 = M_1M_2 = \sqrt{2}$. And now it is obvious, what the remainder between the abscissas of the points M_1 and M_2 is equal to 1.

These considerations resulted us in the following equation for the calculation of the abscissa of the point M_2, that is, $x_{M_2} = x$:

$$x - x^{-1} = 1 \quad \text{or} \quad x^2 - x - 1 = 0. \tag{3.59}$$

This means that the abscissa $x_{M_2} = x$ is a positive root of the famous "golden" algebraic equation:

$$x_{M_2} = \Phi = \frac{1+\sqrt{5}}{2}. \tag{3.60}$$

Thus, a study of the metric properties of the phyllotaxis lattice in Fig.3.12, 3.13 unexpectedly resulted in the golden ratio. And this fact is the *second key outcome of Bodnar's geometry*. This result was used by Bodnar for the detailed study of phyllotaxis phenomenon. By developing this idea, Bodnar concluded that for the mathematical simulation of phyllotaxis phenomenon we need to use a special class of the hyperbolic functions, named *"golden" hyperbolic functions* [20]:

The "golden" hyperbolic sine

$$Gshn = \frac{\Phi^n - \Phi^{-n}}{2} \tag{3.61}$$

The "golden" hyperbolic cosine

$$Gchn = \frac{\Phi^n + \Phi^{-n}}{2}. \tag{3.62}$$

In further, Bodnar found a fundamental connection of the "golden" hyperbolic functions (3.61), (3.62) with Fibonacci numbers:

$$F(2k-1) = \frac{2}{\sqrt{5}} Gch(2k-1) \qquad (3.63)$$

$$F(2k) = \frac{2}{\sqrt{5}} Gsh\, 2k. \qquad (3.64)$$

By using the correlations (3.63), (3.64), Bodnar gave very simple explanation of the "puzzle of phyllotaxis": why Fibonacci numbers occur with such persistent constancy on the surface of phyllotaxis objects. The main reason consists in the fact that the geometry of the "Alive Nature," in particular, geometry of phyllotaxis, is a non-Euclidean geometry; but this geometry is substantially distinguished from Lobachevski's geometry and Minkovsky's four-dimensional world, based on the classical hyperbolic functions (3.1), (3.2). This distinction consists in the fact that the main correlations of this geometry are described with the help of the "golden" hyperbolic functions (3.61) and (3.62), connected with the Fibonacci numbers by the simple relations (3.63) and (3.64).

It is important to emphasize that Bodnar's model of the dynamic symmetry of phyllotaxis objects, illustrated by Fig.3.11, is confirmed brilliantly by real phyllotaxis pictures of botanic objects (see, for example, Fig.3.8, 3.9).

3.7.5. Connection of Bodnar's "golden" hyperbolic functions with the symmetric hyperbolic Fibonacci and Lucas functions

By comparing the expressions for the symmetric hyperbolic Fibonacci and Lucas sine's and cosines, given by the formulas (3.12) – (3.15), with the expressions for Bodnar's "golden" hyperbolic functions, given by the formulas (3.61), (3.62), we can find the following simple correlations between the indicated groups of the formulas:

$$Gsh(x) = \frac{\sqrt{5}}{2} sFs(x) \qquad (3.65)$$

$$Gch(x) = \frac{\sqrt{5}}{2} cFs(x) \qquad (3.66)$$

$$Gsh(x) = 2sLs(x) \qquad (3.67)$$

$$Gch(x) = 2cLs(x) \qquad (3.68)$$

The analysis of these correlations allows to conclude that the "golden" hyperbolic sine and cosine introduced by Oleg Bodnar [20] and the symmetric hyperbolic Fibonacci and Lucas sine's and cosines, introduced by Stakhov and Rozin in [10-12], coincide within constant factors. A question of the use of the "golden" hyperbolic functions or the hyperbolic Fibonacci and Lucas functions for the simulation of phyllotaxis objects has not a particular significance because the final result will be the same: always it will result in the unexpected appearance of the Fibonacci or Lucas numbers on the surfaces of phyllotaxis objects.

3.7.6. Significance of Bodnar's geometry for the development of hyperbolic geometry

The main result of scientific research by Oleg Bodnar consists in the following. Bodnar created a new kind of hyperbolic geometry, which is the basis of the widespread botanical phenomenon of phyllotaxis.

We can do the following important conclusions, which follow from Bodnar's geometry:

1. "Bodnar's geometry" has discovered for modern science the new "hyperbolic world" – the world of phyllotaxis and its geometric secrets. The main feature of this world is the fact that the basic mathematical relations of this world are described by the hyperbolic Fibonacci functions, which are a reason of the appearance of Fibonacci numbers on the surface of phyllotaxis objects.

2. "Bodnar's geometry" has showed that hyperbolic geometry is much more common in the real world than it seemed before. The hyperbolic Fibonacci and Lucas functions, introduced in [10-12], are "natural" functions of Nature. They appear in different botanical structures such, as pine cones, pineapples, cacti, heads of sunflower and so on. Bodnar's geometry is a new hyperbolic geometry of wildlife and this fact is of fundamental importance for the development of modern sciences for the "Living Nature" (biology, botany, physiology, medicine, genetics, and so on).

3. Comparing the classical hyperbolic geometry, created by Nikolay Lobachevski, and the new hyperbolic geometry of phyllotaxis, created by Oleg Bodnar [20], we should indicate the fundamental distinction between them. Lobachevski's geometry is based on the classical hyperbolic

functions (3.1), (3.2), which use "Euler' constant" of e as the base of these functions. The applications of Lobachevski's geometry relate, first of all, to the "Mineral World" and physical phenomena (Einstein's relativity theory, four-dimensional Minkowsky's world, etc.). "Bodnar's geometry" is a hyperbolic geometry of the "Living Nature." It is based on the symmetric hyperbolic Fibonacci functions (3.12), (3.13), which use the golden ratio as the base of these functions. In contrast to the classical hyperbolic functions, the symmetric hyperbolic Fibonacci and Lucas functions own the unique mathematical properties. The first of them is the fact that they possess the recursive properties, similar to Fibonacci numbers. The second property follows from the fundamental mathematical connection of the symmetric hyperbolic Fibonacci and Lucas functions with the "extended" Fibonacci numbers what is the main scientific explanation why the Fibonacci spirals appear on the surface of phyllotaxis objects.

4. The last conclusion relates to the influence of Bodnar's geometry on the future development of hyperbolic geometry. As is known, the hyperbolic geometry is the result of the replacement of the Euclidean 5th postulate (the postulate of parallel lines) to the new postulate - "Lobachevski's postulate." Since then, the non-Euclidean geometries begun to be created by the way of "postulates' replacement" what led to the emergence of new non-Euclidean geometries (for example, elliptic geometry and so on). In Lobachevski's times, only one class of hyperbolic functions, determined by the formulas (3.1), (3.2), was known. It is the use of these functions is always meant, when we speak about the "hyperbolic geometry". Creating a new class of hyperbolic functions, called the hyperbolic Fibonacci and Lucas functions [10-12], became a prerequisite for the creation of a new kind of hyperbolic geometry, called "Bodnar's geometry" [20]. "Bodnar's geometry" has been created by means of the "hyperbolic functions replacement." It discovers a new way of the hyperbolic geometry development: a search for new hyperbolic functions, which can lead to the creation of new hyperbolic geometries. This idea is developing in Chapter 4 and 5.

5. Finally, it is useful to evaluate the hyperbolic Fibonacci and Lucas functions and "Bodnar's geometry" in the terms of *Dirac's Principle of Mathematical Beauty*. The hyperbolic Fibonacci and Lucas functions are

based on the golden ratio, one of the most beautiful mathematical discoveries of ancient science. This remarkable mathematical object is within the *Dirac's Principle of Mathematical Beauty,* and therefore the hyperbolic Fibonacci and Lucas functions and the associated with them "Bodnar's geometry" fully comply with this principle. Bodnar's geometry revealed the secret of the phyllotaxis phenomenon - one of the most amazing phenomena of wildlife. If Nature actually operates on the scenario, suggested by Oleg Bodnar, then we must recognize that Nature is a great mathematician, used the hyperbolic Fibonacci and Lucas functions in its objects during millions years, with moment of the wildlife emergence.

Chapter 4

FIBONACCI AND LUCAS LAMBDA-NUMBERS, "METALLIC MEANS," AND HYPERBOLIC FIBONACCI AND LUCAS LAMBDA FUNCTIONS

4.1. The Fibonacci λ-numbers

4.1.1. A brief history

In the late 20 th and early 21 th centuries, several researchers from different countries –Argentinean mathematician **Vera W. de Spinadel** [52], French mathematician **Midhat Gazale** [53], American mathematician **Jay Kappraff** [54], Russian engineer **Alexander Tatarenko** [55], Armenian philosopher and physicist **Hrant Arakelyan** [56], Russian researcher **Victor Shenyagin** [57], Ukrainian physicist **Nikolai Kosinov** [58], Spanish mathematicians **Sergio Falcon** and **Angel Plaza** [59] and others independently one to another began to study a new class of recursive numerical sequences, which are a generalization of the classical Fibonacci numbers. These numerical sequences led to the discovery of a new class of mathematical constants, called "metallic means" by Vera W. de Spinadel [52].

The interest of many independent researchers from different countries (USA, Canada, Argentina, France, Spain, Russia, Armenia, Ukraine) can not be accidental. This means that the problem of the generalization of Fibonacci numbers and "golden ratio" has matured in modern science.

4.1.2. Definition

Let us give a real number $\lambda > 0$ and consider the following recurrence relation:

$$F_\lambda(n+2) = \lambda F_\lambda(n+1) + F_\lambda(n) \tag{4.1}$$

with the seeds:

$$F_\lambda(0) = 0, F_\lambda(1) = 1. \tag{4.2}$$

The recurrence relation (4.1) for the seeds (4.2) generates an infinite number of new numerical sequences, because every real number $\lambda > 0$ "generates" its own numerical sequence.

Let us consider the partial cases of the recurrence relation (4.1). For the case $\lambda = 1$ the recurrence relation (4.1) and the seeds (4.2) are reduced to the following:

$$F_1(n+2) = F_1(n+1) + F_1(n) \tag{4.3}$$

$$F_1(0) = 0, F_1(1) = 1 \tag{4.4}$$

The recurrence relation (4.3) for the seeds (4.4) generates the classical Fibonacci numbers:

$$0, 1, 1, 2, 3, 5, 8, 13, 21, 34, \ldots \tag{4.5}$$

Based on this fact, we will name a general class of the numerical sequences, generated by the recurrence relation (4.1) for the seeds (4.2), the **Fibonacci λ-numbers**.

For the case $\lambda = 2$ the recurrence relation (4.1) and the seeds (4.2) are reduced to the following:

$$F_2(n+2) = 2F_2(n+1) + F_2(n) \tag{4.6}$$

$$F_2(0) = 0, F_2(1) = 1. \tag{4.7}$$

The recurrence relation (4.6) for the seeds (4.7) generates the so-called **Pell numbers**

$$0, 1, 2, 5, 12, 29, 70, 169, 408, \ldots \tag{4.8}$$

For the cases $\lambda = 3, 4$ the recurrence relation (4.1) and the seeds (4.2) are reduced to the following:

$$F_2(n+2) = 3F_2(n+1) + F_2(n); \ F_2(0) = 0, F_2(1) = 1 \tag{4.9}$$

$$F_2(n+2) = 4F_2(n+1) + F_2(n); \ F_2(0) = 0, F_2(1) = 1. \tag{4.10}$$

4.2. The generalized Cassini's formula for the Fibonacci λ-numbers

4.2.1. The "extended" Fibonacci λ-numbers

The Fibonacci λ-numbers have many remarkable properties, similar to the properties of the classical Fibonacci numbers. It easy to prove that the Fibonacci λ-numbers, as well as the classical Fibonacci numbers, can be "extended" to the negative values of the discrete variable n.

Table 4.1 shows the four "extended" Fibonacci λ-sequences, corresponding to the values $\lambda = 1, 2, 3, 4$.

Table 4.1. The "extended" Fibonacci λ-numbers ($\lambda = 1, 2, 3, 4$)

n	0	1	2	3	4	5	6	7	8
$F_1(n)$	0	1	1	2	3	5	8	13	21
$F_1(-n)$	0	1	-1	2	-3	5	-8	13	-21
$F_2(n)$	0	1	2	5	12	29	70	169	408
$F_2(-n)$	0	1	-2	5	-12	29	-70	169	-408
$F_3(n)$	0	1	3	10	33	109	360	1189	3927
$F_3(-n)$	0	1	-3	10	-33	109	-360	1199	-3927
$F_4(n)$	0	1	4	17	72	305	1292	5473	23184
$F_4(-n)$	0	1	-4	17	-72	305	-1292	5473	-23184

4.2.2. A proof of the generalized Cassini's formula

Let us prove that Cassini's formula (2.21) can be generalized for the case of the Fibonacci λ-numbers. The generalized Cassini formula is of the following form [60]:

$$F_\lambda^2(n) - F_\lambda(n-1) F_\lambda(n+1) = (-1)^{n+1} \qquad (4.11)$$

Let us prove a validity of the formula (4.11) by the induction on n. For the case $n = 1$ the Fibonacci λ-numbers $F_\lambda(n-1), F_\lambda(n), F_\lambda(n+1)$ in the formula (4.11), according to the recurrence relation (4.1), takes the following values: $F_\lambda(0) = 0, F_\lambda(1) = 1, F_\lambda(2) = \lambda$, what implies that the identity (4.11) for the case $n = 1$ is equal:

$$(1)^2 - 0 \times \lambda = (-1)^2. \qquad (4.12)$$

The base of the induction is proved.

We formulate the following inductive assumption. Suppose that the identity (4.11) is valid for the given natural number n, then we prove the validity of the identity (4.11) for the case of $n+1$, that is, the identity

$$F_\lambda^2(n+1) - F_\lambda(n) F_\lambda(n+2) = (-1)^{n+2} \qquad (4.13)$$

is valid too.

In order to prove the identity (4.11), we represent the left-hand part of the identity (4.13) as follows:

$$
\begin{aligned}
F_\lambda^2(n+1) - F_\lambda(n) F_\lambda(n+2) &= F_\lambda^2(n+1) - F_\lambda(n)\left[F_\lambda(n) + \lambda F_\lambda(n+1)\right] = \\
&= F_\lambda^2(n+1) - F_\lambda^2(n) - \lambda F_\lambda(n) F_\lambda(n+1) = \\
&= F_\lambda(n+1)\left[F_\lambda(n+1) - \lambda F_\lambda(n)\right] - F_\lambda^2(n) = F_\lambda(n+1) F_\lambda(n-1) - F_\lambda^2(n)
\end{aligned} \qquad (4.14)
$$

On the other hand, by using (4.11), we can write:

$$F_\lambda(n+1) F_\lambda(n-1) - F_\lambda^2(n) = -(-1)^{n+1}. \qquad (4.15)$$

This means that the identity (4.14) can be rewritten as follows:

$$F_\lambda^2(n+1) - F_\lambda(n) F_\lambda(n+2) = -(-1)^{n+1} = (-1)^{n+2}$$

The identity (4.13) is proved.

4.2.3. The numerical examples

Consider the examples of the validity of the identity (4.11) for the various sequences, shown in Table 4.1. Let us consider the $F_2(n)$-sequence for the case of $n=7$. For this case we should consider the following triple of the Fibonacci 2-numbers $F_2(n)$:

$$F_2(6) = 70, \ F_2(7) = 169, \ F_2(8) = 408.$$

By performing the calculations over them according to (4.11), we get the following result:

$$(169)^2 - 70 \times 408 = 28561 - 28560 = 1,$$

what corresponds to the identity (4.11), because for the case $n=7$ we have:

$$(-1)^{n+1} = (-1)^8 = 1.$$

Now let us consider the $F_3(n)$-sequence from Table 4.1 for the case of $n=6$. For this case, we should chose the following triple of the Fibonacci 3-numbers $F_3(n)$:

$$F_3(5)=109, \; F_3(6)=360, \; F_3(7)=1189.$$

By performing calculations over them according to (4.11), we get the following result:

$$(360)^2-109\times1189=129600-129601=-1,$$

what corresponds to the identity (4.11), because for the case $n=6$ we have

$$(-1)^{n+1}=(-1)^7=-1.$$

Finally, let us consider the $F_4(-n)$-sequence from Table 4.1 for the case of $n=-5$. For this case we should chose the following triple of the Fibonacci 4-numbers $F_4(-n)$:

$$F_4(-4)=-72, \; F_4(-5)=305, \; F_4(-6)=-1292.$$

By performing calculations over them according to (4.11), we get the following result:

$$(305)^2-(-72)\times(-1292)=93025-93024=1,$$

what corresponds to the identity (4.11), because for the case $n=-5$ we have

$$(-1)^{n+1}=(-1)^{-4}=1.$$

Thus, by studying the generalized Cassini formula (4.11) for the Fibonacci λ-numbers, we came to the discovery of an infinite number of integer recurrence sequences in the range from $+\infty$ to $-\infty$, with the following unique mathematical property, expressed by the generalized Cassini formula (4.11), which sounds as follows:

The quadrate of any Fibonacci λ-number $F_\lambda(n)$ are always different from the product of the two adjacent Fibonacci λ-numbers $F_\lambda(n-1)$ and $F_\lambda(n+1)$, which surround the initial Fibonacci λ-number $F_\lambda(n)$, by the number 1; herewith the sign of the difference of 1 depends on the parity of n: if n is even, then the difference of 1 is taken with the sign "minus," otherwise, with the sign "plus."

Until now, we have assumed that only the classical Fibonacci numbers have the unusual property, given by Cassini's formula (2.21). However, as is shown above, a number of such numerical sequences are infinite. All the Fibonacci λ-numbers, generated by the recurrence relation (4.1) for the seeds (4.2), are of a similar property, given by the generalized Cassini's formula (4.11)!

As is well known, a study of integer sequences is the area of number theory. The Fibonacci λ-numbers are integers for the cases $\lambda = 1, 2, 3, \dots$. **Therefore, for many mathematicians in the field of number theory, the existence of the infinite number of the integer sequences, which satisfy to the generalized Cassini's formula (4.11), may be a big surprise.**

4.3. The "Metallic Means"

4.3.1. The algebraic equation for the Fibonacci λ-numbers

Let us represent the recurrence relation (4.1) as follows:

$$\frac{F_\lambda(n+2)}{F_\lambda(n+1)} = \lambda + \frac{1}{\dfrac{F_\lambda(n+1)}{F_\lambda(n)}} \ . \tag{4.16}$$

For the case $n \to \infty$, the expression (4.16) is reduced to the following quadratic equation:

$$x^2 - \lambda x - 1 = 0 . \tag{4.17}$$

with the roots

$$x_1 = \frac{\lambda + \sqrt{4 + \lambda^2}}{2} \quad \text{and} \quad x_2 = \frac{\lambda - \sqrt{4 + \lambda^2}}{2} . \tag{4.18}$$

For the proof of the equation (4.17) we consider auxiliary *point transformation*

$$\bar{s} = f(s) = \lambda + \frac{1}{s} \tag{4.19}$$

where s is an arbitrary real number, different from zero, here at $s \to 0+0$ we have $\bar{s} \to +\infty$, and at $s \to 0-0$ we have $\bar{s} \to -\infty$.

In particular, if we take $s = \dfrac{F_\lambda(n+1)}{F_\lambda(n)}$, than by comparing (4.19) and (4.16),

we get

$$\bar{s} = f(s) = \frac{F_\lambda(n+2)}{F_\lambda(n+1)} \, .$$

Geometrically the fixed point of the transformation (4.19) can be obtained at the intersection of the curve (4.19) with the bisector $\bar{s} = s$, and algebraically they can be obtained as roots of the equation:

$$\lambda + \frac{1}{s} = s \qquad\qquad (4.20)$$

The transformations (4.19) have exactly two fixed points $s^* = x_1, s^{**} = x_2$ of the kind (4.18), and hence they are the roots of the square equation (4.17).

This consideration allows determining two characteristic fixed points – *attractive* and *repelling*. The *attractive point,* denoted by ξ, is a *limiting point* for the iterations $s_k = f^k(s)$, where k is a number of iterations. At $k \to +\infty$, all initial points s belong to any neighbourhood $U(\xi)$ of the point ξ. The *repelling point* is a *limiting point* for the iterations $s_{-k} = f^{-k}(s)$, where k is a number of iterations. At $k \to +\infty$, all initial points s belong to any neighbourhood $U(\xi)$ of the point ξ.

Note that in literature the *attractive* and *repelling* isolated fixed points are called *zero-dimensional attractor* and *zero-dimensional repeller,* respectively.

Let us denote by $f_s'(\xi)$ the first derivative on s of the function $f(s)$ at the fixed point ξ. It is proved in [56-58], that a sufficient condition for the fixed point ξ of the transformation $\bar{s} = f(s)$ to be *attractive* or *repelling* are the following inequalities for the derivative $f_s'(\xi)$, respectively:

$$\left| f_s'(\xi) \right| < 1 \ \text{ or } \ \left| f_s'(\xi) \right| > 1 \qquad\qquad (4.21)$$

This sufficient indication is called *Kenig's theorem* [61 - 63].

Note, that in [61 - 63] the more old terms *"stable* and *unstable fixed points"* are used instead of the terms *attractor* and *repeller.*

Fig.4.1 demonstrates geometric picture for obtaining of the fixed points in the transformation (4.19) for the case $\lambda = 2$.

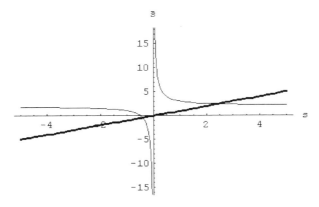

Figure 4.1 A geometric picture for obtaining of the fixed points in the transformation (4.18) for the case $\lambda = 2$.

The direct calculation at $\lambda = 2$ gives the following results. The transformation (4.19) has two fixed points:

1) the point $s^* = 2.41421$ is *attractive*, because $\left| f_s'\left(s^*\right)\right| = 0.171573 < 1$;

2) the point $s^{**} = -0.41421$ is *repelling*, because $\left| f_s'\left(s^{**}\right)\right| = 5.82843 > 1$.

In general case for any fixed λ ($0 < \lambda < +\infty$), as it is stated above, the transformation (4.19) has two fixed points $s^* = x_1$ and $s^{**} = x_2$ of the kind (4.18); at that, because $\left| f_s'(s)\right| = \dfrac{1}{s^2}$, then after simple calculations we get that for the fixed point $s^* = x_1 = \dfrac{\lambda + \sqrt{4 + \lambda^2}}{2}$ we have $\left| f_s'\left(s^*\right)\right| < 1$, that is, the fixed point s^* is *attractive*, and hence

$$\lim_{k \to +\infty} f^k(s) = s^*. \tag{4.22}$$

For the fixed point $s^{**} = x_2 = \dfrac{\lambda - \sqrt{4 + \lambda^2}}{2}$ we have: $\left| f_s'\left(s^{**}\right)\right| > 1$, that is, the fixed point s^{**} is *repelling* and hence

$$\lim_{k \to +\infty} f^{-k}(s) = s^{**}. \tag{4.23}$$

Note that for the first case the initial point s should be chosen for any neighbourhood $U\left(s^*\right)$ of *attractive* fixed point s^*; in the second case the initial

124

point s should be chosen for any neighbourhood $U(s^{**})$ of *repelling* fixed point s^{**}.

The graphs of the functions

$$h^*(\lambda) = \left| f'_s(s^*) \right| = \frac{4}{\left(\lambda + \sqrt{4 + \lambda^2} \right)^2}$$

and

$$h^{**}(\lambda) = \left| f'_s(s^{**}) \right| = \frac{4}{\left(\lambda - \sqrt{4 + \lambda^2} \right)^2}$$

for all $\lambda > 0$ are represented in Fig.4.2 and 4.3. It follows from these figures that $h^*(\lambda) = \left| f'_s(s^*) \right| < 1$ and $h^{**}(\lambda) = \left| f'_s(s^{**}) \right| > 1$.

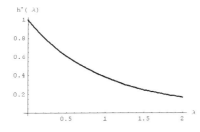

Figure 4.2. Graph of the function
$$h^*(\lambda) = \left| f'_s(s^*) \right|$$

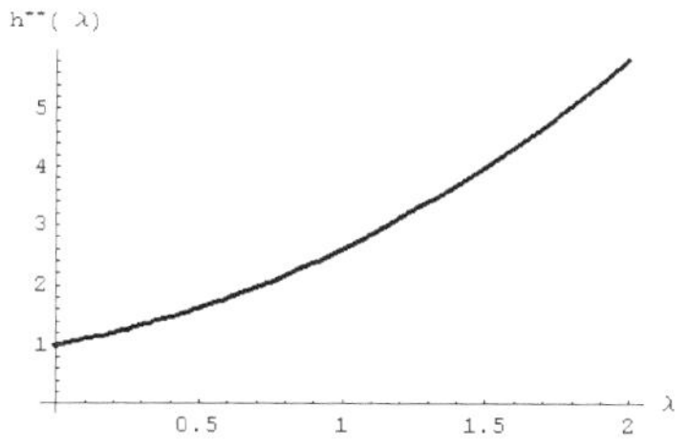

Figure 4.3. Graph of the function
$$h^{**}(\lambda) = \left| f'_s(s^{**}) \right|$$

Then, if we take the ratio $s = \dfrac{F_\lambda(2)}{F_\lambda(1)}$ as the initial value, by virtue of (4.16), we get the following iterations:

$$f^1(s) = \frac{F_\lambda(3)}{F_\lambda(2)}, f^2(s) = \frac{F_\lambda(4)}{F_\lambda(3)}, ..., f^k(s) = \frac{F_\lambda(k+2)}{F_\lambda(k+1)}, \qquad (4.24)$$

Assume that $n = k + 1$, then, taking into consideration (4.24), we get from (4.22):

$$\lim_{n \to +\infty} \frac{F_\lambda(n+1)}{F_\lambda(n)} = s^* = x_1 = \frac{\lambda + \sqrt{4 + \lambda^2}}{2} \qquad . \qquad (4.25)$$

By analogy, if we take into consideration (4.23), we get:

$$\lim_{n \to +\infty} \frac{F_\lambda(-n)}{F_\lambda(-n-1)} = s^{\bullet\bullet} = x_2 = \frac{\lambda - \sqrt{4+\lambda^2}}{2}.$$
(4.26)

4.3.2. The "metallic means" by Vera W. de Spinadel

As mentioned, the equation (4.17) has two roots, a positive root

$$x_1 = \frac{\lambda + \sqrt{4+\lambda^2}}{2}$$
(4.27)

and a negative root

$$x_2 = \frac{\lambda - \sqrt{4+\lambda^2}}{2}$$
(4.28)

If we sum (4.27) and (4.28), we get:

$$x_1 + x_2 = \lambda$$
(4.29)

If we substitute the root (4.27) in place of x in Eq. (4.17), we get the following identity:

$$x_1^2 = \lambda x_1 + 1$$
(4.30)

If we multiply or divide repeatedly all terms of the identity (4.30) by x_1, we get the following identity:

$$x_1^n = \lambda x_1^{n-1} + x_1^{n-2},$$
(4.31)

where $n = 0, \pm 1, \pm 2, \pm 3, \ldots$.

By using this reasoning for the root x_2, we get the following identity for the root x_2:

$$x_2^n = \lambda x_2^{n-1} + x_2^{n-2}$$
(4.32)

Denote the positive root x_1 by Φ_λ, that is,

$$\Phi_\lambda = \frac{\lambda + \sqrt{4+\lambda^2}}{2}$$
(4.33)

Note what for the case $\lambda = 1$ the formula (4.33) is reduced to the formula for the golden ratio:

$$\Phi_1 = \frac{1+\sqrt{5}}{2}.$$
(4.34)

This means that the formula (4.33) gives a wide class of the new mathematical constants, which are a generalization of the golden ratio (4.34).

Basing on this analogy, the Argentinean mathematician **Vera W. de Spinadel** [52] named the mathematical constants (4.33) **metallic means**. If we take $\lambda = 1, 2, 3, 4$ in (4.33), then we get the following mathematical constants having according to Vera de Spinadel the following special names:

$$\Phi_1 = \frac{1+\sqrt{5}}{2} \text{ (the Golden Mean, } \lambda = 1);$$

$$\Phi_2 = 1+\sqrt{2} \text{ (the Silver Mean , } \lambda = 2);$$

$$\Phi_3 = \frac{3+\sqrt{13}}{2} \text{ (the Bronze Mean, } \lambda = 3);$$

$$\Phi_4 = 2+\sqrt{5} \text{ (the Copper Mean, } \lambda = 4).$$

Other metallic means ($\lambda \geq 5$) do not have special names:

$$\Phi_5 = \frac{5+\sqrt{29}}{2}; \quad \Phi_6 = 3+2\sqrt{10}; \quad \Phi_7 = \frac{7+2\sqrt{14}}{2}; \quad \Phi_8 = 4+\sqrt{17}. \tag{4.35}$$

4.3.3. Algebraic properties of the "metallic means"

Let us represent the root x_2 through the "metallic mean" (4.33). After simple transformation of (4.28) we can write the root x_2 as follows:

$$x_2 = \frac{\lambda - \sqrt{4+\lambda^2}}{2} = \frac{\left(\lambda - \sqrt{4+\lambda^2}\right)\left(\lambda + \sqrt{4+\lambda^2}\right)}{2\left(\lambda + \sqrt{4+\lambda^2}\right)} = \frac{-4}{2\left(\lambda + \sqrt{4+\lambda^2}\right)} = -\frac{1}{\Phi_\lambda}. \tag{4.36}$$

Substituting Φ_λ in place of x_1 and $\left(-\dfrac{1}{\Phi_\lambda}\right)$ in place of x_2 in (4.29), we get:

$$\lambda = \Phi_\lambda - \frac{1}{\Phi_\lambda}, \tag{4.37}$$

where Φ_λ is given by (4.33) and $\dfrac{1}{\Phi_\lambda}$ is given by the formula:

$$\frac{1}{\Phi_\lambda} = \frac{-\lambda + \sqrt{4+\lambda^2}}{2}. \tag{4.38}$$

Using the formulas (4.33) and (4.38), we can write the following identity:

$$\Phi_\lambda + \frac{1}{\Phi_\lambda} = \sqrt{4+\lambda^2} \qquad (4.39)$$

Also it is easy to prove the following identity:

$$\Phi_\lambda^n = \lambda\Phi_\lambda^{n-1} + \Phi_\lambda^{n-2}, \qquad (4.40)$$

where $n=0, \pm1, \pm2, \pm3, \dots$.

4.3.4. Two surprising representations of the "metallic means"

For the case $n=2$ the identity (4.40) can be represented in the form:

$$\Phi_\lambda^2 = 1 + \lambda\Phi_\lambda \qquad (4.41)$$

It follows from (4.41) the following representation of the "metallic mean":

$$\Phi_\lambda = \sqrt{1 + \lambda\Phi_\lambda} \qquad (4.42)$$

Substituting $\sqrt{1+\lambda\Phi_\lambda}$ in place of Φ_λ in the right-hand part of (4.42), we get:

$$\Phi_\lambda = \sqrt{1 + \lambda\sqrt{1 + \Phi_\lambda}} \qquad (4.43)$$

Continuing this process ad infinitum, that is, substituting repeatedly $\sqrt{1+\lambda\Phi_\lambda}$ in place of Φ_λ in the right-hand part of (4.43), we get the following surprising representation of the "metallic mean" Φ_λ:

$$\Phi_\lambda = \sqrt{1 + \lambda\sqrt{1 + \lambda\sqrt{1 + \lambda\sqrt{1+\dots}}}} \qquad (4.44)$$

Represent now the identity (4.41) in the form:

$$\Phi_\lambda = \lambda + \frac{1}{\Phi_\lambda} \qquad (4.45)$$

Substituting $\lambda + \frac{1}{\Phi_\lambda}$ in place of Φ_λ in the right-hand part of (4.45), we get:

$$\Phi_\lambda = \lambda + \frac{1}{\lambda + \dfrac{1}{\Phi_\lambda}}. \qquad (4.46)$$

Continuing this process ad infinitum, that is, substituting $\lambda + \frac{1}{\Phi_\lambda}$ repeatedly in place of Φ_λ in the right-hand part of (4.46), we get the following surprising representation of the "metallic mean" Φ_λ:

$$\Phi_\lambda = \lambda + \cfrac{1}{\lambda + \cfrac{1}{\lambda + \cfrac{1}{\lambda + \ldots}}} \qquad (4.47)$$

Note that for the case $\lambda = 1$ the representations (4.44) and (4.47) coincide with the well known representations of the classical golden ratio in the forms:

$$\Phi = \sqrt{1 + \sqrt{1 + \sqrt{1 + \sqrt{1 + \ldots}}}}; \quad \Phi = 1 + \cfrac{1}{1 + \cfrac{1}{1 + \cfrac{1}{1 + \ldots}}} . \qquad (4.48)$$

The representations of the "metallic means" in the forms (4.44) and (4.47), similar to the surprising representations (4.48), are the additional confirmation of the fact that the **"metallic means"** Φ_λ **are the new mathematical constants of mathematics!**

4.4. Gazale's formulas for the Fibonacci and Lucas λ-numbers

4.4.1. Gazale's formula for the Fibonacci λ-numbers

The formulas (4.1), (4.2) define the Fibonacci λ-numbers $F_\lambda(n)$ recursively. However, we can represent the numbers $F_\lambda(n)$ in explicit form through the "metallic mean" Φ_λ. With this purpose, we can represent the Fibonacci λ-numbers $F_\lambda(n)$ by the roots x_1 and x_2 in the form:

$$F_\lambda(n) = k_1 x_1^n + k_2 x_2^n \qquad (4.49)$$

where k_1 and k_2 are constant coefficients, which are the solutions of the following system of the algebraic equations:

$$\begin{cases} F_\lambda(0) = k_1 x_1^0 + k_2 x_2^0 = k_1 + k_2 \\ F_\lambda(1) = k_1 x_1^1 + k_2 x_2^1 = k_1 \Phi_\lambda - k_2 \dfrac{1}{\Phi_\lambda} \end{cases} . \qquad (4.50)$$

Taking into consideration that $F_\lambda(0) = 0$ and $F_\lambda(1) = 1$, we can rewrite the system (4.50) as follows:

$$k_1 = -k_2 \tag{4.51}$$

$$k_1\Phi_\lambda + k_1 \frac{1}{\Phi_\lambda} = k_1\left(\Phi_\lambda + \frac{1}{\Phi_\lambda}\right) = 1. \tag{4.52}$$

Taking into consideration (4.51) and (4.52), we can find the following formulas for the coefficients k_1 and k_2:

$$k_1 = \frac{1}{\sqrt{4+\lambda^2}}; \quad k_2 = -\frac{1}{\sqrt{4+\lambda^2}} \tag{4.53}$$

Taking into consideration (4.53), we can write the formula (4.49) as follows:

$$F_\lambda(n) = \frac{1}{\sqrt{4+\lambda^2}} x_1^n - \frac{1}{\sqrt{4+\lambda^2}} x_2^n = \frac{1}{\sqrt{4+\lambda^2}}\left(x_1^n - x_2^n\right) \tag{4.54}$$

Taking into consideration that $x_1 = \Phi_\lambda$ and $x_2 = -\frac{1}{\Phi_\lambda}$, we can rewrite the formula (4.54) as follows:

$$F_\lambda(n) = \frac{\Phi_\lambda^n - (-1/\Phi_\lambda)^n}{\sqrt{4+\lambda^2}} \tag{4.55}$$

or

$$F_\lambda(n) = \frac{1}{\sqrt{4+\lambda^2}}\left[\left(\frac{\lambda+\sqrt{4+\lambda^2}}{2}\right)^n - \left(\frac{\lambda-\sqrt{4+\lambda^2}}{2}\right)^n\right] \tag{4.56}$$

For the partial case $\lambda = 1$, the formula (4.56) is reduced to the formula:

$$F_1(n) = \frac{1}{\sqrt{5}}\left[\left(\frac{1+\sqrt{5}}{2}\right)^n - \left(\frac{1-\sqrt{5}}{2}\right)^n\right] \tag{4.57}$$

This formula is called *Binet's formula*. This formula was obtained by French mathematician Binet in 1843, although the result was known to Euler, Daniel Bernoulli, and de Moivre more than a century earlier. In particular, de Moivre obtained this formula in 1718.

Note that for the first time the formula (4.55) was deduced by the French mathematician Midhat J. Gazale in the book [53]. This formula sets forth in analytical form an infinite number of the Fibonacci λ-numbers, which for the case $\lambda = 1$ are reduced to Binet's formula (4.57) for the classical Fibonacci numbers. Taking into consideration the uniqueness of the formula (4.55), this formula in Stakhov's article [13] was named *Gazale's formula for the Fibonacci* λ-*numbers* or simply *Gazale's formula.*

4.4.2. Surprising properties of the Fibonacci λ-numbers

Let us find some surprising properties of the Fibonacci λ-numbers by using Gazale's formula (4.55). First of all, compare the Fibonacci λ-numbers $F_\lambda(n)$ and $F_\lambda(-n)$. We can rewrite the formula (4.55) as follows:

$$F_\lambda(n) = \frac{\Phi_\lambda^n - (-1)^n \Phi_\lambda^{-n}}{\sqrt{4+\lambda^2}} \tag{4.58}$$

Let us consider the formula (4.58) for the negative values of *n,* that is,

$$F_\lambda(-n) = \frac{\Phi_\lambda^{-n} - (-1)^{-n} \Phi_\lambda^n}{\sqrt{4+\lambda^2}} \tag{4.59}$$

By comparing the expression (4.58) and (4.59) for the even ($n=2k$) and odd ($n=2k+1$) values of *n,* we can find:

$$\begin{cases} F_\lambda(2k) = -F_\lambda(-2k) \\ F_\lambda(2k+1) = F_\lambda(-2k-1) \end{cases} \tag{4.60}$$

This means what the sequences of the Fibonacci λ-numbers $F_\lambda(n)$ and $F_\lambda(-n)$ in the range $n = 0, \pm 1, \pm 2, \pm 3, \ldots$ are symmetrical sequences relative to $F_\lambda(0) = 0$, if we take into consideration that the Fibonacci λ-numbers $F_\lambda(2k)$ and $F_\lambda(-2k)$ are opposite by sign, that is, the Fibonacci λ-numbers $F_\lambda(n)$ and $F_\lambda(-n)$ are connected by the following simple relation:

$$F_\lambda(-n) = (-1)^{n+1} F_\lambda(n) \tag{4.61}$$

This property of the Fibonacci λ-numbers $F_\lambda(n)$ and $F_\lambda(-n)$ is represented in Table 4.1.

4.4.3. Gazale's formula for the Lucas λ-numbers

Alexey Stakhov in the article [13] has obtained the following result. By analogy to the classical Lucas numbers, we can consider the formula

$$L_\lambda(n) = x_1^n + x_2^n, \tag{4.62}$$

where x_1, x_2 are the roots (4.27), (4.28) of the algebraic equation (4.17).

It is clear that for the case $\lambda = 1$ this formula sets forth the classical Lucas numbers: $2, 1, 3, 4, 7, 11, 18, \ldots$.

Let us assume that for the general case of λ the formula (4.62) sets forth the Lucas λ-numbers. For the given λ, we can find some peculiarities of the Lucas λ-numbers. First of all, by using (4.62), we can calculate the seeds of the Lucas λ-numbers. In fact, for the case $n=0$ and $n=1$ we have, respectively:

$$L_\lambda(0) = x_1^0 + x_2^0 = 1 + 1 = 2 \; ; \tag{4.63}$$

$$L_\lambda(1) = x_1^1 + x_2^1 = \lambda \; . \tag{4.64}$$

By using (4.31) and (4.32), we can represent the formula (4.62) as follows:

$$\begin{aligned} L_\lambda(n) &= x_1^n + x_2^n = \lambda x_1^{n-1} + x_2^{n-2} + \lambda x_1^{n-2} + x_2^{n-2} = \\ &= \lambda\left(x_1^{n-1} + x_2^{n-1}\right) + \left(x_1^{n-2} + x_2^{n-2}\right) \end{aligned} \tag{4.65}$$

Taking into consideration the definition (4.62), we can rewrite (4.65) in the form of the following recurrence relation:

$$L_\lambda(n) = \lambda L_\lambda(n-1) + L_\lambda(n-2) \; . \tag{4.66}$$

It is clear that the recurrence relation (4.66) for the seeds (4.63), (4.64) sets forth **the Lucas λ-numbers**.

If we make the substitution $x_1 = \Phi_\lambda$ and $x_2 = -\dfrac{1}{\Phi_m}$ into the formula (4.62), we can represent the Lucas λ-numbers through the "metallic mean" Φ_λ:

$$L_\lambda(n) = \left[\Phi_\lambda^n + \left(\frac{-1}{\Phi_\lambda}\right)^n \right] \tag{4.67}$$

Although this formula is absent in the book [53], according to the suggestion of Alexey Stakhov [13] the formula (4.67) was named *Gazale's formula for the Lucas λ-numbers* after Midchat Gazale, who first introduced the formula (4.55).

We can rewrite the formula (4.67) as follows:

$$L_\lambda(n) = \Phi_\lambda^n + (-1)^n \Phi_\lambda^{-n} \; . \tag{4.68}$$

Let us consider the formula (4.68) for the negative values of n, that is,

$$L_m(-n) = \Phi_m^{-n} + (-1)^{-n} \Phi_m^n \; . \tag{4.69}$$

By comparing the expression (4.68) and (4.69) for the even ($n=2k$) and odd ($n=2k+1$) values of n, we can write:

$$\begin{cases} L_\lambda(2k) = L_\lambda(-2k) \\ L_\lambda(2k+1) = -L_\lambda(-2k-1) \end{cases}. \tag{4.70}$$

This means what the sequences of the Lucas λ-numbers in the range $n = 0, \pm 1, \pm 2, \pm 3, \ldots$ is symmetrical sequence in respect to $L_\lambda(0) = 2$, excepting that the Lucas λ-numbers $L_\lambda(2k+1)$ and $L_\lambda(-2k-1)$ are opposite by sign, that is, the Lucas λ-numbers $L_\lambda(2k+1)$ and $L_\lambda(-2k-1)$ are connected with the following simple relation:

$$L_\lambda(-n) = (-1)^n L_\lambda(n). \tag{4.71}$$

This property of the Lucas λ-numbers $L_\lambda(n)$ and $L_\lambda(-n)$ is demonstrated in Table 4.2.

Table 4.2. The "extended" Lucas λ-numbers

n	0	1	2	3	4	5	6	7	8
$L_1(n)$	2	1	3	4	7	11	18	29	47
$L_1(-n)$	2	1	−3	4	−7	11	−18	29	−47
$L_2(n)$	2	2	6	14	34	82	198	478	1154
$L_2(-n)$	2	−2	6	−14	34	−82	198	−478	1154
$L_3(n)$	2	3	11	36	119	393	1298	4287	14159
$L_3(-n)$	2	−3	11	−36	119	−393	1298	−4287	14159
$L_4(n)$	2	4	18	76	322	1364	5778	24476	103682
$L_4(-n)$	2	−4	18	−76	322	−1364	5778	−24476	103682

Note what for the case $\lambda = 1$ the Lucas λ-numbers coincide with the classical Lucas numbers, but for the case $\lambda = 2$ with the Pell-Lucas numbers.

4.5. The "Golden" Fibonacci λ-goniometry

4.5.1. A definition of the hyperbolic Fibonacci and Lucas λ-functions

First of all, let us explain the term of *goniometry* used in this book. As is known, a goniometry is a part of geometry, which sets forth relations between trigonometric functions. In further, we use in place of trigonometric functions the *hyperbolic Fibonacci and Lucas λ-functions*, introduced in [13].

In order to determine these functions, we use the identities (4.61) and (4.71) to represent Gazale's formulas (4.55), (4.68) in the following form [13,14]:

$$F_\lambda(n) = \begin{cases} \dfrac{\Phi_\lambda^n - \Phi_\lambda^{-n}}{\sqrt{4+\lambda^2}} & \text{for} \quad n = 2k \\[3mm] \dfrac{\Phi_\lambda^n + \Phi_\lambda^{-n}}{\sqrt{4+\lambda^2}} & \text{for} \quad n = 2k+1 \end{cases} \tag{4.72}$$

$$L_\lambda(n) = \begin{cases} \Phi_\lambda^n - \Phi_\lambda^{-n} & \text{for} \quad n = 2k+1 \\ \Phi_\lambda^n + \Phi_\lambda^{-n} & \text{for} \quad n = 2k \end{cases} \tag{4.73}$$

Comparing Gazale's formulas (4.72) and (4.73) with the classical hyperbolic functions

$$sh(x) = \frac{e^x - e^{-x}}{2}; \quad ch(x) = \frac{e^x + e^{-x}}{2} \tag{4.74}$$

we can see that the formulas (4.72) and (4.73) are similar to the formulas (4.74) by their mathematical structure. This similarity became a reason to introduce a general class of hyperbolic functions called in [13,14] the **hyperbolic Fibonacci and Lucas λ-functions:**

Hyperbolic Fibonacci λ-sine and λ-cosine

$$sF_\lambda(x) = \frac{\Phi_\lambda^x - \Phi_\lambda^{-x}}{\sqrt{4+\lambda^2}} = \frac{1}{\sqrt{4+\lambda^2}}\left[\left(\frac{\lambda+\sqrt{4+\lambda^2}}{2}\right)^x - \left(\frac{\lambda+\sqrt{4+\lambda^2}}{2}\right)^{-x}\right] \tag{4.75}$$

$$cF_\lambda(x) = \frac{\Phi_\lambda^x + \Phi_\lambda^{-x}}{\sqrt{4+\lambda^2}} = \frac{1}{\sqrt{4+\lambda^2}}\left[\left(\frac{\lambda+\sqrt{4+\lambda^2}}{2}\right)^x + \left(\frac{\lambda+\sqrt{4+\lambda^2}}{2}\right)^{-x}\right] \tag{4.76}$$

Hyperbolic Lucas λ-sine and λ-cosine

$$sL_\lambda(x) = \Phi_\lambda^x - \Phi_\lambda^{-x} = \left(\frac{\lambda+\sqrt{4+\lambda^2}}{2}\right)^x - \left(\frac{\lambda+\sqrt{4+\lambda^2}}{2}\right)^{-x} \tag{4.77}$$

$$cL_\lambda(x) = \Phi_\lambda^x + \Phi_\lambda^{-x} = \left(\frac{\lambda+\sqrt{4+\lambda^2}}{2}\right)^x + \left(\frac{\lambda+\sqrt{4+\lambda^2}}{2}\right)^{-x}, \tag{4.78}$$

where x is continuous variable and $\lambda > 0$ is a given positive real number.

It is easy to see that the functions (4.75), (4.76) and (4.77), (4.78) are connected by very simple relations:

$$sF_\lambda(x) = \frac{sL_\lambda(x)}{\sqrt{4+\lambda^2}}; \quad cF_\lambda(x) = \frac{cL_\lambda(x)}{\sqrt{4+\lambda^2}}. \tag{4.79}$$

This means that the hyperbolic Lucas λ-functions (4.77) and (4.78) coincide with the hyperbolic Fibonacci λ-functions (4.75) and (4.76) to within of the constant coefficient $\frac{1}{\sqrt{4+\lambda^2}}$.

4.5.2. An uniqueness of the hyperbolic Fibonacci and Lucas λ-functions

It should be noted the following unique properties of the hyperbolic Fibonacci and Lucas λ- function (4.75)–(4.78):

1. The hyperbolic Fibonacci and Lucas λ-functions are, on the one hand, the generalization of the classical hyperbolic functions (4.74), but on the other hand, the generalization of the symmetric hyperbolic functions Fibonacci and Lucas functions (3.12)–(3.15), which are a special case of the functions (4.75) – (4.78) for the case $\lambda = 1$.

2. Their uniqueness consists in the fact that they, on the one hand, posses all hyperbolic properties, inherent for the classical hyperbolic functions (4.74). On the other hand, they posses all recursive properties, inherent for the symmetric hyperbolic Fibonacci and Lucas functions (3.12)–(3.15).

3. The next unique feature of the functions (4.75) – (4.78) is that the general formulas (4.75) – (4.75) define theoretically infinite number of the new classes of hyperbolic functions, because every positive real number $\lambda > 0$ sets forth a new, previously unknown class of hyperbolic functions.

4. The next unique feature of the functions (4.75) – (4.78) is their unique connection with the "extended" Fibonacci and Lucas λ-numbers, defined by Gazale's formulas (4.72), (4.73). This connection is determined identically by the following relations:

$$\begin{cases} F_\lambda(n) & = \begin{cases} sF_\lambda(n), & n = 2k \\ cF_\lambda(n), & n = 2k+1 \end{cases} \\ L_\lambda(n) & = \begin{cases} cL_\lambda(n), & n = 2k \\ sL_\lambda(n), & n = 2k+1 \end{cases} \end{cases}$$ (4.79)

4.5.3. The graphs of the hyperbolic Fibonacci and Lucas λ-functions

The graphs of the hyperbolic Fibonacci and Lucas λ-functions (Fig.4.4) are similar to the graphs of the symmetric hyperbolic Fibonacci and Lucas functions (see Fig.3.6 and Fig.3.7).

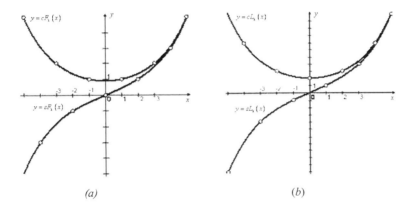

(a)   *(b)*

Figure 4.4. A graph of the hyperbolic Fibonacci (a) and Lucas (b) λ-functions .

It is necessity to note that in the point $x=0$, the hyperbolic Fibonacci λ-cosine $cF_\lambda(x)$ (4.76) takes the value $cF_\lambda(0) = \dfrac{2}{\sqrt{4+\lambda^2}}$, and the hyperbolic Lucas cosine $cL_\lambda(x)$ (4.78) takes the value $cL_\lambda(0) = 2$. It is also important to note that the Fibonacci λ-numbers $F_\lambda(n)$ with the even values of $n = 0, \pm2, \pm4, \pm6, \ldots$ are "inscribed" into the graph of the hyperbolic Fibonacci λ-sine $sF_\lambda(x)$ in the "even" discrete points $x = 0, \pm2, \pm4, \pm6, \ldots$ and the Fibonacci λ-numbers $F_\lambda(n)$ with the odd values of $n = \pm1, \pm3, \pm5, \ldots$ are "inscribed" into the hyperbolic Fibonacci λ-cosine $cF_\lambda(x)$ in the "odd" discrete points $x = \pm1, \pm3, \pm5, \ldots$.

On the other hand, the Lucas λ-numbers $L_\lambda(n)$ with the even values of n are "inscribed" into the graph of the hyperbolic Lucas λ-cosine $cL_\lambda(x)$ in the "even" discrete points $x = 0, \pm2, \pm4, \pm6, \ldots$, and the Lucas λ-numbers $L_\lambda(n)$ with the odd values of n are "inscribed" into the graph of the hyperbolic Lucas λ-sine $sL_\lambda(x)$ in the "odd" discrete points $x = \pm1, \pm3, \pm5, \ldots$.

By analogy with the symmetric hyperbolic Fibonacci and Lucas functions (3.12) - (3.15), we can introduce other kinds of the hyperbolic Fibonacci and Lucas λ-functions, in particular, hyperbolic Fibonacci and Lucas λ-tangents and λ-cotangents, λ-secants and λ-cosecants and so on.

4.6. The partial cases of the hyperbolic Fibonacci and Lucas λ-functions

4.6.1. The "golden," "silver," "bronze," and "copper" hyperbolic functions

Let us consider the partial cases of the hyperbolic Fibonacci and Lucas λ-functions (4.75)-(4.78) for the different values of λ.

For the case $\lambda=1$, the *"golden mean"* (4.34) is a base of the hyperbolic Fibonacci and Lucas 1-functions, which are reduced to the symmetric hyperbolic Fibonacci and Lucas functions (3.12) - (3.15). Therefore in further we will name the functions (3.12) - (3.15) the *"golden" hyperbolic Fibonacci and Lucas functions*.

For the case $\lambda=2$, the *"silver mean"* $\Phi_2 = 1+\sqrt{2}$ is a base of the new class of hyperbolic functions. We will name them the *"silver" hyperbolic Fibonacci and Lucas functions*:

$$sF_2(x) = \frac{\Phi_2^x - \Phi_2^{-x}}{\sqrt{8}} = \frac{1}{2\sqrt{2}}\left[\left(1+\sqrt{2}\right)^x - \left(1+\sqrt{2}\right)^{-x}\right] \qquad (4.80)$$

$$cF_2(x) = \frac{\Phi_2^x + \Phi_2^{-x}}{\sqrt{8}} = \frac{1}{2\sqrt{2}}\left[\left(1+\sqrt{2}\right)^x + \left(1+\sqrt{2}\right)^{-x}\right] \qquad (4.81)$$

$$sL_2(x) = \Phi_2^x - \Phi_2^{-x} = \left(1+\sqrt{2}\right)^x - \left(1+\sqrt{2}\right)^{-x} \qquad (4.82)$$

$$cL_2(x) = \Phi_2^x + \Phi_2^{-x} = \left(1+\sqrt{2}\right)^x + \left(1+\sqrt{2}\right)^{-x} \qquad (4.83)$$

For the case $\lambda=3$ the *"bronze mean"* $\Phi_3 = \dfrac{3+\sqrt{13}}{2}$ is a base of the new class of hyperbolic functions. We will name them the *"bronze"* hyperbolic Fibonacci and Lucas functions:

$$sF_3(x) = \frac{\Phi_3^x - \Phi_3^{-x}}{\sqrt{13}} = \frac{1}{\sqrt{13}}\left[\left(\frac{3+\sqrt{13}}{2}\right)^x - \left(\frac{3+\sqrt{13}}{2}\right)^{-x}\right] \tag{4.84}$$

$$cF_3(x) = \frac{\Phi_3^x + \Phi_3^{-x}}{\sqrt{13}} = \frac{1}{\sqrt{13}}\left[\left(\frac{3+\sqrt{13}}{2}\right)^x + \left(\frac{3+\sqrt{13}}{2}\right)^{-x}\right] \tag{4.85}$$

$$sL_3(x) = \Phi_3^x - \Phi_3^{-x} = \left(\frac{3+\sqrt{13}}{2}\right)^x - \left(\frac{3+\sqrt{13}}{2}\right)^{-x} \tag{4.86}$$

$$cL_3(x) = \Phi_3^x + \Phi_3^{-x} = \left(\frac{3+\sqrt{13}}{2}\right)^x + \left(\frac{3+\sqrt{13}}{2}\right)^{-x} \tag{4.87}$$

For the case $\lambda=4$ the *"copper mean"* $\Phi_4 = 2+\sqrt{5}$ is a base of the new class of hyperbolic functions. We will name them the *"copper"* hyperbolic Fibonacci and Lucas functions:

$$sF_4(x) = \frac{\Phi_4^x - \Phi_4^{-x}}{2\sqrt{5}} = \frac{1}{2\sqrt{5}}\left[\left(2+\sqrt{5}\right)^x - \left(2+\sqrt{5}\right)^{-x}\right] \tag{4.88}$$

$$cF_4(x) = \frac{\Phi_4^x + \Phi_4^{-x}}{2\sqrt{5}} = \frac{1}{2\sqrt{5}}\left[\left(2+\sqrt{5}\right)^x + \left(2+\sqrt{5}\right)^{-x}\right] \tag{4.89}$$

$$sL_4(x) = \Phi_4^x - \Phi_4^{-x} = \left(2+\sqrt{5}\right)^x - \left(2+\sqrt{5}\right)^{-x} \tag{4.90}$$

$$cL_4(x) = \Phi_4^x + \Phi_4^{-x} = \left(2+\sqrt{5}\right)^x + \left(2+\sqrt{5}\right)^{-x} \tag{4.91}$$

Note that a list of these functions can be continued ad infinitum. Note that, because $\lambda>0$ is a positive real number, the number of the hyperbolic Fibonacci and Lucas λ-functions is equal to the number of positive real numbers.

4.6.2. Comparison of the classical hyperbolic functions with the hyperbolic Lucas λ-functions

Let us compare the hyperbolic Lucas λ-functions (4.77) and (4.78) with the classical hyperbolic functions (4.74). It is easy to prove [54] that for the case

$$\Phi_\lambda = \frac{\lambda + \sqrt{4 + \lambda^2}}{2} = e \qquad (4.92)$$

the hyperbolic Lucas λ-functions (4.77) and (4.78) coincide with the classical hyperbolic functions (4.74) to within of the constant coefficient $\frac{1}{2}$, that is,

$$sh(x) = \frac{sL_\lambda(x)}{2} \quad \text{and} \quad ch(x) = \frac{cL_\lambda(x)}{2}. \qquad (4.93)$$

By using (4.92), after simple transformations we can calculate the value of λ_e, for which the expressions (4.93) are valid:

$$\lambda_e = e - \frac{1}{e} = 2sh(1) \approx 2.35040238. \qquad (4.94)$$

Thus, according to (4.93) the classical hyperbolic functions (4.74) are a partial case of the hyperbolic Lucas λ-functions for the case (4.94).

4.7. The most important formulas for the hyperbolic Fibonacci and Lucas λ-functions

4.7.1. The relations between the "metallic means" and the "golden mean"

We emphasize once again that the number of hyperbolic Fibonacci and Lucas λ-functions, given by (4.75)-(4.78), can be extended to infinity. It is important to emphasize that they have a relation to some well-known numerical sequences, in particular, to the Fibonacci, Lucas and Pell numbers. These functions keep all the important properties of the classical hyperbolic functions and the symmetric hyperbolic Fibonacci and Lucas functions (3.12)-(3.15); moreover, on the one hand, they have the recurrent properties, similar to Fibonacci and Lucas λ-numbers and, on the other hand, the hyperbolic properties, similar to classical hyperbolic functions, which are a particular case of the hyperbolic Lucas λ-functions.

The "metallic means" (4.33), which are a generalization of the classic "golden ratio," are the bases of these functions. Let's start with the relations between the "metallic means" and the "golden ratio" (Table 4.3).

Table 4.3. Comparative table for the Golden Ratio and Metallic Means

The Golden Ratio ($\lambda = 1$)	The Metallic Means ($\lambda > 0$)
$\Phi = \dfrac{1 + \sqrt{5}}{2}$	$\Phi_\lambda = \dfrac{\lambda + \sqrt{4 + \lambda^2}}{2}$
$\Phi = \sqrt{1 + \sqrt{1 + \sqrt{1 + \sqrt{\ldots}}}}$	$\Phi_\lambda = \sqrt{1 + \lambda\sqrt{1 + \lambda\sqrt{1 + \lambda\sqrt{\ldots}}}}$
$\Phi = 1 + \cfrac{1}{1 + \cfrac{1}{1 + \cfrac{1}{1 + \ldots}}}$	$\Phi_\lambda = \lambda + \cfrac{1}{\lambda + \cfrac{1}{\lambda + \cfrac{1}{\lambda + \ldots}}}$
$\Phi^n = \Phi^{n-1} + \Phi^{n-2} = \Phi \times \Phi^{n-1}$	$\Phi_\lambda^n = \lambda\Phi_\lambda^{n-1} + \Phi_\lambda^{n-2} = \Phi_\lambda \times \Phi_\lambda^{n-1}$
$F(n) = \dfrac{\Phi^n - (-1)^n \Phi^{-n}}{\sqrt{5}}$	$F_\lambda(n) = \dfrac{\Phi_\lambda^n - (-1)^n \Phi_\lambda^{-n}}{\sqrt{4 + \lambda^2}}$
$L(n) = \Phi^n + (-1)^n \Phi^{-n}$	$L_\lambda(n) = \Phi_\lambda^n + (-1)^n \Phi_\lambda^{-n}$
$sFs(x) = \dfrac{\Phi^x - \Phi^{-x}}{\sqrt{5}}$	$sF_\lambda(x) = \dfrac{\Phi_\lambda^x - \Phi_\lambda^{-x}}{\sqrt{4 + \lambda^2}}$
$cFs(x) = \dfrac{\Phi^x + \Phi^{-x}}{\sqrt{5}}$	$cF_\lambda(x) = \dfrac{\Phi_\lambda^x + \Phi_\lambda^{-x}}{\sqrt{4 + \lambda^2}}$
$sLs(x) = \Phi^x - \Phi^{-x}$	$sL_\lambda(x) = \Phi_\lambda^x - \Phi_\lambda^{-x}$
$cLs(x) = \Phi^x + \Phi^{-x}$	$cL_\lambda(x) = \Phi_\lambda^x + \Phi_\lambda^{-x}$

4.7.2. The recurrence properties

Let us prove now some recurrence properties for the hyperbolic Fibonacci and Lucas λ-functions.

Theorem 4.1. The following relations, which are similar to the recurrence relation for the Fibonacci λ-numbers $F_\lambda(n+2) = \lambda F_\lambda(n+1) + F_\lambda(n)$ are valid for the hyperbolic Fibonacci and Lucas λ-functions:

$$sF_\lambda(x+2) = \lambda cF_\lambda(x+1) + sF_\lambda(x) \qquad (4.95)$$

$$cF_\lambda(x+2) = \lambda sF_\lambda(x+1) + cF_\lambda(x) . \qquad (4.96)$$

Proof:

$$sF_\lambda(x+2) = \lambda cF_\lambda(x+1) + sF_\lambda(x) = \lambda\frac{\Phi_\lambda^{x+1} + \Phi_\lambda^{-x-1}}{\sqrt{4+\lambda^2}} + \frac{\Phi_\lambda^x - \Phi_\lambda^{-x}}{\sqrt{4+\lambda^2}} =$$
$$= \frac{\Phi_\lambda^x (\lambda\Phi + 1) - \Phi_\lambda^{-x}(1 - \lambda\Phi_\lambda^{-1})}{\sqrt{4+\lambda^2}} \qquad (4.97)$$

Since $\lambda\Phi_\lambda + 1 = \Phi_\lambda^2$ and $1 - \Phi_\lambda^{-1} = \Phi_\lambda^{-2}$, we can represent (4.97) as follows:

$$\lambda c F_\lambda (x+1) + s F_\lambda (x) = \frac{\Phi_\lambda^{x+2} - \Phi_\lambda^{-x-2}}{\sqrt{4+\lambda^2}} = s F_\lambda (x+2),$$

what proves the identity (4.95).

By analogy we can prove the identity (4.96).

We proved above the generalized Cassini's formula for the Fibonacci λ-numbers. given by (4.11). This formula can be generalized on the continuous domain for the case of hyperbolic Fibonacci λ-functions.

Theorem 4.2 (a generalization of Cassini's formula for the continuous domain). The following relations, which are similar to the generalized Cassini's formula for the Fibonacci λ-numbers $F_\lambda^2(n) - F_\lambda(n-1)F_\lambda(n+1) = (-1)^{n+1}$ are valid for the hyperbolic Fibonacci λ-functions:

$$\left[s F_\lambda (x) \right]^2 - c F_\lambda (x+1) c F_\lambda (x-1) = -1 \tag{4.98}$$

$$\left[c F_\lambda (x) \right]^2 - s F_\lambda (x+1) s F_\lambda (x-1) = 1. \tag{4.99}$$

Proof:

$$\left[s F_\lambda (x) \right]^2 - c F_\lambda (x+1) c F_\lambda (x-1) =$$

$$= \left(\frac{\Phi_\lambda^x - \Phi_\lambda^{-x}}{\sqrt{4+\lambda^2}} \right)^2 - \frac{\Phi_\lambda^{x+1} + \Phi_\lambda^{-x-1}}{\sqrt{4+\lambda^2}} \times \frac{\Phi_\lambda^{x-1} + \Phi_\lambda^{-x+1}}{\sqrt{4+\lambda^2}} =$$

$$= \frac{\Phi_\lambda^{2x} - 2 + \Phi_\lambda^{-2x} - \left(\Phi_\lambda^{2x} + \Phi_\lambda^2 + \Phi_\lambda^{-2} + \Phi_\lambda^{-2x} \right)}{4+\lambda^2} = \tag{4.100}$$

$$= \frac{-2 - \left(\Phi_\lambda^2 + \Phi_\lambda^{-2} \right)}{4+\lambda^2}.$$

Using the formula (4.73), for the case $n=2$ we can write:

$$L_\lambda(2) = \Phi_\lambda^2 + \Phi_\lambda^{-2}. \tag{4.101}$$

Using the recurrence relation (4.66) and the initial conditions (4.63), (4.64), we can represent the Lucas λ-number $L_\lambda(2)$ as follows:

$$L_\lambda(2) = \lambda L_\lambda(1) + L_\lambda(0) = \lambda \times \lambda + 2 = 2 + \lambda^2. \tag{4.102}$$

Taking into consideration (4.101) and (4.102), we can conclude that (4.98) holds for any real number $\lambda > 0$.

By analogy, we can also prove the identity (4.99). Theorem 4.2 is proved.

Note that Theorems 4.1 and 4.2 are the examples of the so-called recurrence properties of hyperbolic Fibonacci λ-functions.

4.7.3. The hyperbolic properties

We now prove some hyperbolic properties of the Fibonacci and Lucas λ-functions (4.75) - (4.78).

We start with the **parity properties**, which are valid for the classical hyperbolic functions. As is known, the hyperbolic cosine is an even function, and the hyperbolic sine is an odd function. We prove that the hyperbolic Fibonacci and Lucas λ-functions also have this important property.

Theorem 4.3. Hyperbolic Fibonacci and Lucas λ-sines are odd functions, and hyperbolic Fibonacci and Lucas λ-cosines are even functions.

Proof:

$$sF_\lambda(-x) = \frac{\Phi_\lambda^{-x} - \Phi_\lambda^{x}}{\sqrt{4+\lambda^2}} = -\frac{\Phi_\lambda^{x} - \Phi_\lambda^{-x}}{\sqrt{4+\lambda^2}} = -sF_\lambda(x), \qquad (4.103)$$

which implies that the hyperbolic Fibonacci λ-sine is an odd function.

On the other hand, the hyperbolic Fibonacci cosine is an even function, because

$$cF_\lambda(-x) = \frac{\Phi_\lambda^{-x} + \Phi_\lambda^{x}}{\sqrt{4+\lambda^2}} = \frac{\Phi_\lambda^{x} + \Phi_\lambda^{-x}}{\sqrt{4+\lambda^2}} = cF_\lambda(x). \qquad (4.104)$$

By analogy, we can prove:

$$sL_\lambda(-x) = -sL_\lambda(x); \ cL_\lambda(-x) = cL_\lambda(x) \qquad (4.105)$$

Theorem 4.4. The following relations, similar to the identity $[ch(x)]^2 - [sh(x)]^2 = 1$ for the classical hyperbolic functions, are valid for the hyperbolic Fibonacci and Lucas λ-functions:

$$[cF_\lambda(x)]^2 - [sF_\lambda(x)]^2 = \frac{4}{4+\lambda^2} \qquad (4.106)$$

$$[cL_\lambda(x)]^2 - [sL_\lambda(x)]^2 = 4. \qquad (4.107)$$

Proof:
We can prove the identity (4.106) as follows:

$$[cF_\lambda(x)]^2 - [sF_\lambda(x)]^2 = \left(\frac{\Phi_\lambda^{x} + \Phi_\lambda^{-x}}{\sqrt{4+\lambda^2}}\right)^2 - \left(\frac{\Phi_\lambda^{x} - \Phi_\lambda^{-x}}{\sqrt{4+\lambda^2}}\right)^2 =$$

$$= \frac{\Phi_\lambda^{2x} + 2 + \Phi_\lambda^{-2x} - \Phi_\lambda^{2x} + 2 - \Phi_\lambda^{-2x}}{4+\lambda^2} = \frac{4}{4+\lambda^2}$$

142

The identity (4.107) can be easy to prove, if we use the relations (4.79) linking the hyperbolic Fibonacci λ-functions with the hyperbolic Lucas λ-functions. Substituting in place of the hyperbolic Fibonacci λ-functions $cF_\lambda(x)$ и $sF_\lambda(x)$ their representations through the hyperbolic Lucas λ-functions, given (4.79), we get:

$$\left[\frac{cL_\lambda(x)}{\sqrt{4+\lambda^2}}\right]^2 - \left[\frac{sL_\lambda(x)}{\sqrt{4+\lambda^2}}\right]^2 = \frac{4}{4+\lambda^2}. \tag{4.108}$$

After performing simple algebraic transformations in (4.108), we get the identity (4.107). Theorem 4.4 is proved.

Theorem 4.5. The following relation, similar to the identity $ch(x+y)=ch(x)ch(y)+sh(x)sh(y)$ for the classical hyperbolic functions, is valid for the hyperbolic Fibonacci λ- functions:

$$\frac{2}{\sqrt{4+\lambda^2}}cF_\lambda(x+y) = cF_\lambda(x)cF_\lambda(y)+sF_\lambda(x)sF_\lambda(y) \tag{4.109}$$

Proof:

$$cF_\lambda(x)cF_\lambda(y)+sF_\lambda(x)sF_\lambda(y) =$$
$$= \frac{\Phi_\lambda^x+\Phi_\lambda^{-x}}{\sqrt{4+\lambda^2}}\times\frac{\Phi_\lambda^y+\Phi_\lambda^{-y}}{\sqrt{4+\lambda^2}}+\frac{\Phi_\lambda^x-\Phi_\lambda^{-x}}{\sqrt{4+\lambda^2}}\times\frac{\Phi_\lambda^y-\Phi_\lambda^{-y}}{\sqrt{4+\lambda^2}} =$$
$$= \frac{\Phi_\lambda^{x+y}+\Phi_\lambda^{x-y}+\Phi_\lambda^{-x+y}+\Phi_\lambda^{-x-y}+\Phi_\lambda^{x+y}-\Phi_\lambda^{x-y}-\Phi_\lambda^{-x+y}+\Phi_\lambda^{-x-y}}{4+\lambda^2} = \tag{4.110}$$
$$= \frac{2\left(\Phi_\lambda^{x+y}+\Phi_\lambda^{-x-y}\right)}{\sqrt{4+\lambda^2}\times\sqrt{4+\lambda^2}} = \frac{2}{\sqrt{4+\lambda^2}}cF_\lambda(x+y)$$

Theorem 4.6. The following identity, similar to the identity $ch(x-y)=ch(x)ch(y)-sh(x)sh(y)$ for the classical hyperbolic functions, is valid for hyperbolic Fibonacci λ-functions:

$$\frac{2}{\sqrt{4+\lambda^2}}cF_\lambda(x-y) = cF_\lambda(x)cF_\lambda(y)-sF_\lambda(x)sF_\lambda(y). \tag{4.111}$$

The proof is similar to Theorem 4.5.

We assume without proof the following theorems.

Theorem 4.7. The following identities, similar to the identity $ch(2x) = [ch(x)]^2 + [sh(x)]^2$ for the classical hyperbolic functions, are valid for the hyperbolic Fibonacci and Lucas λ-functions:

$$\frac{2}{\sqrt{4+\lambda^2}} cF_\lambda(2x) = \left[cF_\lambda(x)\right]^2 + \left[sF_\lambda(x)\right]^2 \tag{4.112}$$

$$2cL_\lambda(2x) = \left[cL_\lambda(x)\right]^2 + \left[sL_\lambda(x)\right]^2. \tag{4.113}$$

Theorem 4.8. The following identities, similar to the identity $sh(2x) = 2sh(x)ch(x)$ for the classical hyperbolic functions, are valid for the hyperbolic Fibonacci and Lucas λ-functions:

$$\frac{1}{\sqrt{4+m^2}} sF_\lambda(2x) = sF_\lambda(x)cF_\lambda(x) \tag{4.114}$$

$$sL_\lambda(2x) = sL_\lambda(x)cL_\lambda(x). \tag{4.115}$$

Theorem 4.9. The following formulas, similar to Moivre's formulas $[ch(x) \pm sh(x)]^n = ch(nx) \pm sh(nx)$ for the classical hyperbolic functions, are valid for the hyperbolic Fibonacci and Lucas λ-functions:

$$\left[cF_\lambda(x) \pm sF_\lambda(x)\right]^n = \left(\frac{2}{\sqrt{4+m^2}}\right)^{n-1} \left[cF_\lambda(nx) \pm sF_\lambda(nx)\right]; \tag{4.116}$$

$$\left[cL_\lambda(x) \pm sL_\lambda(x)\right]^n = 2^{n-1}\left[cL_\lambda(nx) \pm sL_\lambda(nx)\right]. \tag{4.117}$$

Table 4.4 contains the basic formulas, connecting the hyperbolic Fibonacci λ-functions with the classical hyperbolic functions. These formulas give the hyperbolic properties of the hyperbolic Fibonacci λ-functions.

Note that the table of the similar formulas for the hyperbolic Lucas λ-functions could be obtained from Table 4.4, if we use the relations (4.79), which connect the hyperbolic Lucas λ-functions with the hyperbolic Fibonacci λ-functions.

Table 4.4. Hyperbolic properties of the hyperbolic Fibonacci λ-functions

Formulas for the classical hyperbolic functions	Formulas for the hyperbolic Fibonacci λ-functions
$sh(x)=\dfrac{e^{x}-e^{-x}}{2};ch(x)=\dfrac{e^{x}+e^{-x}}{2}$	$sF_{\lambda}(x)=\dfrac{\Phi_{\lambda}^{x}-\Phi_{\lambda}^{-x}}{\sqrt{4+\lambda^{2}}};cF_{\lambda}(x)=\dfrac{\Phi_{\lambda}^{x}+\Phi_{\lambda}^{-x}}{\sqrt{4+\lambda^{2}}}$
$sh(x+2)=2sh(1)ch(x+1)+sh(x)$ $ch(x+2)=2sh(1)sh(x+1)+ch(x)$	$sF_{\lambda}(x+2)=\lambda cF_{\lambda}(x+1)+sF_{\lambda}(x)$ $cF_{\lambda}(x+2)=\lambda sF_{\lambda}(x+1)+cF_{\lambda}(x)$
$sh^{2}(x)-ch(x+1)ch(x-1)=-ch^{2}(1)$ $ch^{2}(x)-sh(x+1)sh(x-1)=ch^{2}(1)$	$\left[sF_{\lambda}(x)\right]^{2}-cF_{\lambda}(x+1)cF_{\lambda}(x-1)=-1$ $\left[cF_{\lambda}(x)\right]^{2}-sF_{\lambda}(x+1)sF_{\lambda}(x-1)=1$
$ch^{2}(x)-sh^{2}(x)=1$	$\left[cF_{\lambda}(x)\right]^{2}-\left[sF_{\lambda}(x)\right]^{2}=\dfrac{4}{4+\lambda^{2}}$
$sh(x+y)=sh(x)ch(x)+ch(x)sh(x)$ $sh(x-y)=sh(x)ch(x)-ch(x)sh(x)$	$\dfrac{2}{\sqrt{4+\lambda^{2}}}sF_{\lambda}(x+y)=sF_{\lambda}(x)cF_{\lambda}(x)+cF_{\lambda}(x)sF_{\lambda}(x)$ $\dfrac{2}{\sqrt{4+\lambda^{2}}}sF_{\lambda}(x-y)=sF_{\lambda}(x)cF_{\lambda}(x)-cF_{\lambda}(x)sF_{\lambda}(x)$
$ch(x+y)=ch(x)ch(x)+sh(x)sh(x)$ $ch(x-y)=ch(x)ch(x)-sh(x)sh(x)$	$\dfrac{2}{\sqrt{4+\lambda^{2}}}cF_{\lambda}(x+y)=cF_{\lambda}(x)cF_{\lambda}(x)+sF_{\lambda}(x)sF_{\lambda}(x)$ $\dfrac{2}{\sqrt{4+\lambda^{2}}}cF_{\lambda}(x-y)=cF_{\lambda}(x)cF_{\lambda}(x)-sF_{\lambda}(x)sF_{\lambda}(x)$
$ch(2x)=2sh(x)ch(x)$	$\dfrac{1}{\sqrt{4+\lambda^{2}}}cF_{\lambda}(2x)=sF_{\lambda}(x)cF_{\lambda}(x)$
$\left[ch(x)\pm sh(x)\right]^{n}=ch(nx)\pm sh(nx)$	$\left[cF_{\lambda}(x)\pm sF_{\lambda}(x)\right]^{n}=\left(\dfrac{2}{\sqrt{4+\lambda^{2}}}\right)^{n-1}\left[cF_{\lambda}(nx)\pm sF_{\lambda}(nx)\right]$

4.8. Conclusions

1. A beauty of the formulas presented in Tables 4.3, 4.4 are charming. This gives a right to suppose that *Dirac's "Principle of Mathematical Beauty* is applicable fully to the metallic means (4.33) and hyperbolic Fibonacci and Lucas λ-functions (4.75) - (4.78). And this, in its turn, gives a hope that the above mathematical results can be used as effective models of many phenomena in theoretical natural sciences.

2. The main results of the works [9-14] is the introduction of the new class of hyperbolic functions, the *hyperbolic Fibonacci and Lucas functions*, based on the *golden ratio* [9-12], and *hyperbolic Fibonacci and Lucas λ-functions* ($\lambda>0$ is a given real number) based on the *metallic means* [13,14]. These new classes of hyperbolic functions are similar to the

classical hyperbolic functions and retain all their useful mathematical properties (*hyperbolic properties*). Besides, they are a generalization of the classical Fibonacci and Lucas numbers and Fibonacci and Lucas λ-numbers, which coincide with the *hyperbolic Fibonacci and Lucas functions* and the *hyperbolic Fibonacci and Lucas λ-functions* for discrete values of continues variable $x = 0, \pm 1, \pm 2, \pm 3, \ldots$, and retain all their useful mathematical properties (*recursive properties*).

Chapter 5

HILBERT'S FOURTH PROBLEM: PSEUDO-SPHERICAL SOLUTION

5.1. Euclid's Fifth Postulate and Lobachevski's geometry

On February 23, 1826 on the meeting of the Mathematics and Physics Faculty of Kazan University the Russian mathematician **Nikolay Lobachevski** (1792 -1856) had proclaimed on the creation of the new geometry, named *imaginary geometry*. This geometry was based on the traditional Euclid's postulates, excepting Euclid's Fifth Postulate about parallels. New Fifth Postulate about parallels was formulated by Lobachevski as follows:

"At the plane through a point outside a given straight line, we can conduct two and only two straight lines parallel to this line, as well as an endless set of straight lines, which do not overlap with this line and are not parallel to this line, and the endless set of straight lines, intersecting the given straight line."

For the first time, the new geometry was outlined by Lobachevski in 1829 in the article *About the Foundations of Geometry* in the magazine *Kazan Bulletin*.

Independently on Lobachevski, the Hungarian mathematician **Janos Bolyai** (1802-1860) came to such ideas. He published his work *Appendix* three years later Lobachevski (1832). Also the prominent German mathematician **Carl Friedrich Gauss** (1777-1855) came to the same ideas. After his death some unpublished sketches on the non-Euclidean geometry were found.

Figure 5.1. Nikolay Lobachevski (1792 -1856); Carl Friedrich Gauss (1777-1855); Janos Bolyai (1802-1860)

Lobachevski's geometry got a full recognition and wide distribution 12 years after his death, when there is became clear that scientific theory, built on the basis of a system of axioms, is considered to be fully completed only when the *system of axioms* meets three conditions: *independence, consistency* and *completeness*. Lobachevski's geometry satisfies these conditions. Finally this became clear in 1868 when the Italian mathematician **Eugenio Beltrami** (1835-1900) in his memoirs *The Experience of the Non-Euclidean Geometry Interpretation* showed that in Euclidean space at pseudo-spherical surfaces geometry of Lobachevski's plane arises, if we take geodesic lines as straight lines.

Later the German mathematician **Felix Christian Klein** (1849-1925) and the French mathematician **Henri Poincare** (1854-1912) proved a consistency of Non-Euclidean geometry, by means of the construction of corresponding models of Lobachevski's plane. The interpretation of Lobachevski's geometry on the surfaces of Euclidean space contributed to general recognition of Lobachevski's ideas.

The creation of *Riemannian geometry* by **Georg Friedrich Bernhard Riemann** (1826-1866), became the main outcome of such Non-Euclidean approach. The Riemannian geometry developed a mathematical doctrine about geometric space, a notion of differential and a distance between elements of diversity and a doctrine about curvature.

The introduction of the *generalized Riemannian spaces*, whose particular cases are *Euclidean space* and *Lobachevski's space*, and the so-called *Riemannian geometry*, opened new ways in the development of geometry. They found their applications in physics (theory of relativity) and other branches of theoretical natural sciences.

Lobachevski's geometry also is called *hyperbolic geometry* because it is based on the *hyperbolic functions,* introduced in 18th century by the Italian mathematician **Vincenzo Riccati** (1707-1775).

The most famous classical interpretations of **Lobachevski's** plane with the *Gaussian curvature K* <0 , are the following:

-**Beltrami's** interpretation on a disk;

- **Poincare's** interpretation on a disk.

-**Klein's** interpretation at a half-plane and other.

Lobachevsky metric is called the metric form

$$(ds)^2 = R^2 \left[(du)^2 + sh^2(u)(dv)^2 \right] \tag{5.1}$$

with the Gaussian curvature $K = -\dfrac{1}{R^2} < 0$, where the variables (u, v) belong to the half-plane

$$\Pi^+ : (U, V), 0 < U < +\infty, -\infty < V < +\infty. \tag{5.2}$$

Here *ds* is called an *arc length* and $sh(u) = \dfrac{e^u - e^{-u}}{2}$ is a *hyperbolic sine.*

Lobachevski's geometry has remarkable applications in many fields of modern natural sciences. This concerns not only applied aspects (cosmology, electrodynamics, plasma theory), but, first of all, it concerns the most fundamental sciences and their foundation - mathematics (number theory, theory of automorphic functions created by **A. Poincare,** geometry of surfaces and so on).

Since on the closed surfaces of the negative Gaussian curvature, *Lobachevski's geometry* is fulfilled and Lobachevski's plane is universal covering for these surfaces, it is very fruitful to study various objects (dynamical systems with continuous and discrete time, layers, fabrics and so on), defined on these surfaces. By developing this idea, we can raise these objects to the level of universal covering, which is replenished by the absolute ("infinity"), and further we can study smooth topological properties of these objects by using the notion of the *absolute.*

Samuil Aranson studied this problem about four decades. The works [64-69],[74], written by Samuil Aranson with co-authors, give a presentation about these results and research methods. Aranson's DrSci dissertation *"Global problems of qualitative theory of dynamic systems on surfaces"* (1990) is devoted to this theme.

5.2.   Hilbert's Fourth Problem

5.2.1. A brief history

In the lecture *"Mathematical Problems"* [6], presented at the Second International Congress of Mathematicians (Paris, 1900), **David Hilbert** (1862-1943) had formulated his famous 23 mathematical problems. These problems determined considerably the development of the 20th century mathematics. This lecture is a unique phenomenon in the mathematics history and in mathematical literature.

Figure 5.2. David Hilbert (1862-1943)

The Russian translation of Hilbert's lecture and its comments are given in the book [7]. In Hilbert's original work [6], Hilbert's Fourth Problem is called as *"Problem of the straight line as the shortest distance between two points."* In [6] this problem has been formulated as follows:

"Another problem relating to the foundations of geometry is this: If from among the axioms necessary to establish ordinary Euclidean geometry, we exclude the axiom of parallels, or assume it as not satisfied, but retain all other axioms, we obtain, as is well known, the geometry of Lobachevski (hyperbolic geometry). We may therefore say that this is a geometry standing next to Euclidean geometry...

The more general question now arises: Whether from other suggestive standpoints geometries may not be devised which, with equal right, stand next to Euclidean geometry"...

The theorem of the straight line as the shortest distance between two points and the essentially equivalent theorem of Euclid about the sides of a triangle, play an important part not only in number theory but also in the theory of surfaces and in the calculus of variations. For this reason, and because I believe that the thorough investigation of the conditions for the validity of this theorem will throw a new light upon the idea of distance, as well as upon other elementary ideas, e. g., upon the idea of the plane, and the possibility of its definition by means of the idea of the straight line, the construction and systematic treatment of the geometries here possible seem to me desirable."

Hilbert's citations contain the formulation of very important mathematical problem, which touches foundation of geometry, number theory, the theory of surfaces and the calculus of variations. Hilbert's Fourth Problem is of fundamental interest not only for mathematics, but also for all theoretical natural sciences: are there non-Euclidean geometries, which are next to Euclidean geometry and are interesting from the *"other suggestive standpoints"*?

If we consider this problem in the context of theoretical natural sciences, then Hilbert's Fourth Problem puts forward the aim searching for NEW HYPERBOLIC WORLDS OF NATURE, which are close to Euclidean geometry and reflect some new properties of Nature's structures and phenomena..

Hilbert considers *Lobachevski's geometry* and *spherical geometry* as the nearest to Euclidean geometry. As it is noted in Wikipedia [8], *"in mathematics, Hilbert's Fourth Problem in the 1900 "Hilbert problems" was a foundational question in geometry. In one statement derived from the original, it was to find geometries whose axioms are closest to those of Euclidean geometry if the ordering and incidence axioms are retained, the congruence axioms weakened, and the equivalent of the parallel postulate omitted."*

In mathematical literature Hilbert's Fourth Problem is sometimes considered as formulated **very vague** what makes difficult its final solution. As it is noted in Wikipedia [8], *"the original statement of Hilbert, however, has also been judged too vague to admit a definitive answer."* In [70] American geometer **Herbert Busemann** analyzes the whole range of issues, related to Hilbert's

Fourth Problem, and also concludes that the question, related to this problem, is unnecessarily broad.

Unfortunately, the attempts solving Hilbert's Fourth Problem, made by German mathematician **Herbert Hamel** (1901) and later by the Soviet mathematician **Alexey Pogorelov** [71] (1974), have not led to significant progress, as pointed out in Wikipedia's articles [8, 72]. In order to learn more about an axiomatic approach to solving the Hilbert's Fourth Problem by the famous Ukrainian geometer **A.V. Pogorelov,** see also Aranson's article [73].

In Wikipedia's article [72], the status of the problem is formulated as *"too vague to be stated resolved or not"* and **Pogorelov's** solution [71] even is not mentioned.

The very cautious point of view on Pogorelov's solution of Hilbert's Fourth Problem is presented in the remarkable book [75]. Thus, from the standpoint of modern mathematical community, Hilbert's blame consists in the fact that he formulated this problem not clearly enough and this is the main reason, why Hilbert's Fourth Problem is not resolve until now.

In spite of critical attitude of mathematicians to Hilbert's Fourth Problem, we should emphasize a great importance of this problem for mathematics and theoretical natural sciences. Without doubts, Hilbert's intuition led him to the conclusion that *Lobachevski's geometry, spherical geometry* and *Minkovski's geometry* do not exhaust all possible variants of non-Euclidean geometries. Hilbert's Fourth Problem directs researchers at searching of new non-Euclidean geometries, which are close to the traditional Euclidean geometry.

5.2.2. From the "game of postulates" to the "game of functions"

According to [76], a cause of the difficulties, associated with the solution of Hilbert's Fourth Problem, lies elsewhere. All the known attempts to solve this problem (Herbert Hamel, Alexey Pogorelov) were in the traditional framework of the so-called "game of postulates" [76]

This "game" started from the works by **Nikolay Lobachevski** and **Janos Bolyai**, when Euclid's 5th postulate was replaced by the opposite one. This was the most major step in the development of the non-Euclidean geometry, which led to **"Lobachevski's geometry."** This geometry changed the traditional geometric ideas and is known as *hyperbolic geometry.* This name highlights the fact that this geometry is based on the classical hyperbolic functions (3.1), (3.2). And this is one of the "key" ideas of "Lobachevski's geometry."

It is important to emphasize that the very name of "hyperbolic geometry" contains another approach to the solution of Hilbert's Fourth Problem: searching for the new classes of "hyperbolic functions," which can be the basis for other hyperbolic geometries. Every new class of the hyperbolic functions generates a new variant of the "hyperbolic geometry." By analogy with the "game of postulates" this approach to solving of Hilbert's Fourth Problem can be named the "game of functions" [76].

5.2.3. Hyperbolic Fibonacci and Lucas functions and "Bodnar's geometry

In this connection, the introduction of the new class of hyperbolic functions, based on the golden ratio [9-12], and following from them new geometric theory of phyllotaxis (*Bodnar's geometry*) [20] are of principal importance for the development of hyperbolic geometry, because it shows on the existence of the new hyperbolic geometries in surrounding us world. **Alexey Stakhov** in [13,14] gave a wide generalization of the symmetric hyperbolic Fibonacci and Lucas functions [10-12] and developed the so-called *hyperbolic Fibonacci and Lucas λ-functions.*

It is proved in [13, 14] the existence of infinite variants of new hyperbolic functions, which can be the base for new hyperbolic geometries. The main purpose of this chapter is to develop this idea, that is, to create the new hyperbolic geometries, based on the hyperbolic Fibonacci and Lucas λ-functions [13,14], and to get a solution of Hilbert's Fourth Problem. This study can be considered as the original solution (the *pseudo-spherical solution*) of *Hilbert's Fourth Problem* based on the "*metallic means*" [52].

5.3. Basic notions and concepts

5.3.1. Preliminary information

Lobachevski's classical metric (5.1) can be obtained, when considering the upper half of the two-sheeted hyperboloid (pseudo sphere of the "radius" R):

$$M^2 : Z^2 - X^2 - Y^2 = R^2 ; Z \geq R > 0, \tag{5.3}$$

153

embedded in three-dimensional pseudo-Euclidean space (X,Y,Z) with Minkowski's metrics

$$(dl)^2 = (dZ)^2 - (dX)^2 - (dY)^2.$$

Here dl is the arc element in the space (X,Y,Z), with the next parameterization of the surface (5.2) in the form:

$$M^2 : X = Rsh(u)\cos(v),\ Y = Rsh(u)\sin(v),\ Z = Rch(u), \tag{5.4}$$

where the half-plane (5.2) $\Pi^+ : (U,V), 0 < U < +\infty, -\infty < V < +\infty$ is the domain of existence of curvilinear coordinates $(u,\ v)$. With this parameterization of M^2 for the metric (5.1) we have: $(ds)^2 = -(dl)^2$.

In the special theory of relativity (STR) we use the following coordinate system: the spatial coordinates $X,\ Y$, the time coordinate $Z = c_0 t > 0$, where c_0 is the light velocity in vacuum, t is a time.

Fig. 5.3 gives a visual representation of the surface M^2. The surface M^2 belongs to the so-called time-homothetic domain, bounded by the upper half of the *isotropic* (in other terminology, *light*) cone K^2: $Z^2 - X^2 - Y^2 = 0$, $Z > 0$.

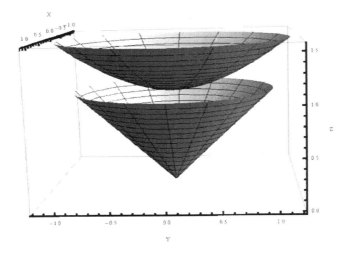

Figure 5.3. The surface M^2

Note that the surface M^2 is considered as the upper half of the two-sheeted hyperboloid, embedded into Minkowski's three-dimensional space. From this

154

point of view, the surface M^2 is considered as an open two-dimensional manifold. All the considered below parametric representation of this surface in those or other curvilinear coordinates are nothing as covering of the manifold M^2 by the different sets of the cards Σ with recalculation from one card to another card, while in this situation, each such card completely covers all the two-dimensional manifolds M^2.

In further, each card will have an independent life in isolation from the surface M^2 as interpretation of those or other Lobachevski's model with specifically introduced metrics (Lobachevski's plane with Lobachevski's metric, Poincare's model on the disk, Klein's model on the half-plane), an infinite set of models of the authors of this article, which preserve invariance of half-plane.

Each such model with specifically introduced metrics has its own individual unique geometric and differential properties and requires of separate investigation.

At the same time when we convert one model into another model, these properties are interpreted differently. In this case, the following questions appear: whether the Gaussian curvature remains constant, how geodesic lines, angles, squares of figures in the particular model are changing, what are movements, compressions, conversions, which are not movements, how we should study dynamical systems, as the one-parameter groups of transformations with continuous and discrete time and so on. In this case, a study of the absolute arithmetic properties for such models, for the limiting continuation of such transformations on the *absolute*, are important from theoretical and practical point of view.

Of particular interest is the study of such models as the universal ramified or non-ramified covering spaces $\overline{M}$, when we present the closed orientable and non-orientable surface M of Euler's negative characteristic $\chi(M)$ in the form of factor $\overline{M}/G$ of the covering spaces $\overline{M}$ for the discrete groups of the transformations G.

5.3.2. The notion of the *absolute*

In our study, the important role plays the concept of the *absolute*, replenishing the space $\overline{M}$. These issues are studied in many works of **Samuil Aranson** (see, for example, the works [64-69],[74]).

In order to introduce the concept of the *absolute* for the following non-Euclidean metric forms, we pre-equip the entire plane or three-dimensional space with Euclidean metrics.

Then we further equip with the non-Euclidean metric form a certain area of the Euclidean plane or a surface in three-dimensional Euclidean space. We will call the absolute of non-Euclidean metric form the boundary ∂D of the domain of definition D of this form, such that at approaching to ∂D "inside" of the domain D, the non-Euclidean form degenerates.

Below we consider three non-Euclidean metric forms of the Gaussian curvature $K = -1$, which are the interpretation of Lobachevski's geometry.

1). For Poincare's metric form

$$(ds)^2 = \frac{4\left[(dx)^2 + (dy)^2\right]}{\left(1 - x^2 - y^2\right)^2}, \quad E = G = \frac{4}{\left(1 - x^2 - y^2\right)^2}, F = 0,$$

given on the disc $D: x^2 + y^2 < 1$ of the Euclidean plane (x, y), the absolute is the so-called "infinity circle" $\partial D = (x^2 + y^2 = 1)$.

At approaching "from within" the area $D: x^2 + y^2 < 1$ along of any ray to the *"infinity circle"* $\partial D = (x^2 + y^2 = 1)$ we get:

$$\lim (E) = \lim (G) = +\infty, \ F = 0,$$

that is, Poincare's metric form degenerates.

2). For Lobachevski's metric form

$$(ds)^2 = (du)^2 + sh^2(u)(dv)^2, E = G = sh^2(u) > 0, F = 0,$$

given on the half-plane $D: u > 0, -\infty < v < +\infty$ of the Euclidean plane (u, v), the absolute ∂D has the following form:

$$\partial D = (u = 0, -\infty < v < +\infty) \ \bigcup (u = +\infty, -\infty < v < +\infty),$$

where $(u = 0, -\infty < v < +\infty)$ - v is the axis of the Euclidean plane (u, v), $(u = +\infty, -\infty < v < +\infty)$ is an *"infinitely distant line."*

At approaching "from within" the area $D: u > 0, -\infty < v < +\infty$ along of any ray to the v-axis $(u = 0, -\infty < v < +\infty)$ we get:

$$\lim (E) = \lim (G) = sh^2(0) = 0, F = 0,$$

and hence Lobachevski's metric form degenerates.

156

A similar situation is obtained also at approaching "from within" the area $D : u > 0, -\infty < v < +\infty$ along of any ray to an *"infinitely distant line"* $(u = +\infty, -\infty < v < +\infty)$, since then

$$\lim (E) = \lim (G) = sh^2(+\infty) = +\infty, \ F = 0,$$

that is, Lobachevski's metric form again degenerates.

Note, that the **classical** metric form, used by Lobachevski, had the following form: $(ds)^2 = (dx)^2 + ch^2\left(\frac{x}{R}\right)(dy)^2$, where the Gaussian curvature $K = -\frac{1}{R^2} < 0$. This form is obtained from the form $(ds)^2 = R^2\left[(du)^2 + sh^2(u)(dv)^2\right]$ of the same Gaussian curvature $K = -\frac{1}{R^2} < 0$ when replacing the parameterization (5.4) of the pseudo-sphere M^2 of the kind (5.3) on the parameterization of the kind:

M^2: $X = Rch\left(\frac{x}{R}\right)sh\left(\frac{y}{R}\right)$, $Y = Rsh\left(\frac{x}{R}\right)$, $Z = Rch\left(\frac{x}{R}\right)ch\left(\frac{y}{R}\right)$. Then, the following one-to-one connection arises between the variables (u,v) and (x,y):

$$sh(u) \cos(v) = ch\left(\frac{x}{R}\right)sh\left(\frac{y}{R}\right), \ sh(u) \sin(v) = sh\left(\frac{x}{R}\right), \ ch(u) = ch\left(\frac{x}{R}\right)ch\left(\frac{y}{R}\right).$$

Hence we obtain the following:

$$-(dl)^2 = (ds)^2 = R^2\left[(du)^2 + sh^2(u)(dv)^2\right],$$

where $(dl)^2 = (dZ)^2 - (dX)^2 - (dY)^2$ is Minkovski's metrics in the space (X, Y, Z).

3). **For Minkovski's metric form**

$$(dl)^2 = (dZ)^2 - (dX)^2 - (dY)^2,$$

where dl is an element of arc, given on the pseudo sphere $D : Z^2 - X^2 - Y^2 = 1, Z \geq 1$ in the Euclidean space (X, Y, Z); here the *absolute* is the *"infinitely distant circumference"* $\partial D = (X^2 + Y^2 = +\infty, Z = +\infty)$, which "belongs" to the upper half $Z = \sqrt{X^2 + Y^2} > 0$ of the *light cone* $K^2 : Z^2 - X^2 - Y^2 = 0$.

Proof. According to [5], the pseudo sphere $D : Z^2 - X^2 - Y^2 = 1, Z \geq 1$ admits a parameterization

$$X = \frac{2x}{1 - (x^2 + y^2)}, \ Y = \frac{2y}{1 - (x^2 + y^2)}, \ Z = \frac{1 + (x^2 + y^2)}{1 - (x^2 + y^2)},$$

where $\quad -(dl)^2 = (ds)^2 = \dfrac{4\left[(dx)^2 + (dy)^2\right]}{\left(1 - x^2 - y^2\right)^2}$.

Poincare's metric form on the disc is the following: $0 \le x^2 + y^2 < 1$.

Since for Poincare's metric form, the *absolute* is the circumference $x^2 + y^2 = 1$ and because $\quad X^2 + Y^2 = \dfrac{4(x^2 + y^2)}{\left[1 - (x^2 + y^2)\right]^2}$, $Z = \dfrac{1 + x^2 + y^2}{1 - (x^2 + y^2)}$, then at approaching

$x^2 + y^2 \to 1$ "from within" the area $0 \le x^2 + y^2 < 1$ we get $X^2 + Y^2 \to +\infty, Z \to +\infty$.

This means that the *"infinitely distant circumference"* $\partial D = (X^2 + Y^2 = +\infty, Z = +\infty)$ is the *absolute* for Minkovski's metric form.

5.3.3. Two-parametric family of linear transformations

Let us pass to the specific results, obtained by the authors and related to Hilbert's Fourth Problem, and their interpretation by using the notions of the "golden ratio" and "metallic proportions" [52].

By direct computation, it is easy to find that the surface M^2 given (5.3) is also invariant at the parameterization

$$\begin{cases} X = Rsh[u(u',v')]\cos[v(u',v')] \\ Y = Rsh[u(u',v')]\sin[v(u',v')]) \\ Z = Rch[u'(u,v)] \end{cases} \qquad (5.5)$$

where $u = u(u',v'), v = v(u',v')$ are smooth functions, $\dfrac{\partial(u,v)}{\partial(u',v')} \neq 0$.

However, for such arbitrary parameterization, the certain geometric structures and differential properties at one-to-one mapping of the surface M^2 can be violated, a priori.

Here and below (unless otherwise stated) we restrict ourselves by the two-parametric family of the linear transformations (5.5) of the half-plane Π^+, for the conditions, when the change of variables (u,v) according to the transformation (5.5) into the variables (u',v') are fulfilled by the rule:

$$f : \begin{cases} f : u = u(u',v') = \alpha \bullet u' \\ v = v(u',v') = \beta \bullet v' \\ \dfrac{\partial(u,v)}{\partial(u',v')} = \alpha\beta \neq 0 \end{cases}, \tag{5.6}$$

where α, β are some real numbers satisfying to the conditions:

$$0 < \alpha, \beta < +\infty \tag{5.7}$$

Thus, in this situation, the transformation (5.6) is the introduction of new coordinates in the half-plane $\varPi^+ : 0 < U < +\infty, -\infty < V < +\infty$, such, when the new coordinates (u',v') everywhere in the half-plane $\varPi^+$ can be expressed through the old coordinates and vice versa. From the viewpoint of Riemannian geometry, the two-parametric transformation (5.6) has their special specific properties.

Further on the half-plane $\varPi^+$ for any fixed positive value of $R>0$ we will compare (unless otherwise stated), two metric forms of the kind:

1). Lobachevski's classical metric form

$$\begin{cases} (ds)^2 = R^2 \left[(du)^2 + sh^2(u)(dv)^2 \right] \\ E = E(u,v) = R^2 > 0, F = F(u,v) = 0, G = G(u,v) = R^2 sh^2(u) > 0 \\ EG - F^2 = R^2 sh^2(u) > 0 \end{cases} \tag{5.8}$$

2). Two-parametric metric form

$$\begin{cases} (ds)^2 = R^2 \left[\alpha^2 (du')^2 + \beta^2 sh^2(\alpha u')(dv')^2 \right] \\ E' = E'(u',v') = R^2 \alpha^2 > 0, F' = F'(u',v') = 0, G' = G'(u',v') = R^2 \beta^2 sh^2(\alpha u') > 0 \\ E'G' - (F')^2 = R^2 \alpha^2 \beta^2 sh^2(\alpha u') > 0 \end{cases} \tag{5.9}$$

The metric form (5.8) is converted into the form (5.9) under the action of diffeomorphism (5.6) for each value of the real numbers α, β, satisfying to the condition (5.7). Here ds, ds' are **elements of arc lengths**, E, F, G, E', F', G' are the coefficients of metric forms.

Metric forms in terms of tensor analysis are symmetric covariant tensor field of the rank two on a smooth manifold. Through this manifold, the scalar product on the tangent space, the length of curves, angles between curves, squares and so on are given.

159

To identify the specific properties of the transformations (5.6) and their effect on the metric forms, pre-recall some basic concepts, related to the nonsingular quadratic metric forms of internal geometry and their transformations induced by diffeomorphisms (see, for example, [5]).

5.3.4. Isometric mapping and equivalence, isometry, and conformal metric forms

First of all, we will make some changes in some definitions and concepts, because different sources give different interpretations of these concepts.

Suppose further that, unless otherwise stated, we consider the half-plane

$$\varPi^+ : 0 < U < +\infty, -\infty < V < +\infty,$$

where three objects are given:

1). Diffeomorphism

$$f: \ u = u(u',v') \ , \ v = v(u',v'), \frac{\partial(u,v)}{\partial(u',v')} \neq 0. \tag{5.10}$$

2). Metric form

$$(ds)^2 = E(u,v)(du)^2 + 2F(u,v)dudv + G(u,v)(dv)^2, \tag{5.11}$$

where the coefficients satisfy to the non-equalities: $E > 0, G > 0, EG - F^2 > 0$

3). Metric form

$$(ds')^2 = E'(u',v')(du')^2 + 2F'(u',v')du'dv' + G'(u',v')(dv')^2, \tag{5.12}$$

where the coefficients satisfy to the non-equalities: $E' > 0, G' > 0, E'G' - (F')^2 > 0$.

Definition 5.1. We say that the diffeomorphism (5.10) is an isometric map if, under the influence of this diffeomorphism, the metric form (5.11) is converted to the metric form (5.12), while the lengths of the elements satisfy the condition:

$$ds[u(u'v'), v(u',v')] = ds'(u',v'). \tag{5.13}$$

The metric forms (5.11) and (5.12), satisfying to the condition (5.13) is called **isometrically equivalent**.

Thus, under the action of isometric mapping, the elements of arc lengths remain the same, although the metric forms in the variables (u,v) and (u',v') may have different forms, and therefore may not preserve the same angles between the arcs.

Under the effect of the isometric mapping (5.10), the metric form (5.11) is converted into the metric form (5.12) of the following form:

$$\left[\begin{array}{l} (ds')^2 = E(u',v')(du')^2 + 2F'(u',v')\,du'dv' + G(u',v')(dv')^2 \\ \begin{pmatrix} E'(u',v') \\ F'(u',v') \\ G'(u',v') \end{pmatrix} \equiv \begin{pmatrix} a_1^2 & 2a_1b_1 & b_1^2 \\ a_1a_2 & a_1b_2+a_2b_1 & b_1b_2 \\ a_2^2 & 2a_2b_2 & b_2^2 \end{pmatrix} \begin{pmatrix} E(u,v) \\ F(u,v) \\ G(u,v) \end{pmatrix} \end{array} \right. \tag{5.14}$$

where

$$u = u(u',v'),\ v = v(u',v'),\ a_1 = \frac{\partial u}{\partial u'},\ a_2 = \frac{\partial u}{\partial v'},\ b_1 = \frac{\partial v}{\partial u'},\ b_2 = \frac{\partial v}{\partial v'}. \tag{5.15}$$

Definition 5.2. We say that the diffeomorphism (5.10) is an **isometry**, if under the action of the diffeomorphism (5.10) the metric form (5.11) is converted into the metric form (5.12) for the variables (u',v') having the same form as the metric form (5.11) for the variables (u,v), that is, for the case of the isometry the coefficients of the metrical forms (5.12) and (5.11) satisfy to the following identities:

$$\left\{ \begin{array}{l} (ds)^2 = E(u,v)(du)^2 + 2F(u,v)\,dudv + G(u,v)(dv)^2 \\ (ds')^2 = E'(u',v')(du')^2 + 2F'(u',v')\,dudv + G(u',v')(dv')^2 \\ E'(u',v') \equiv E(u,v),\ F'(u',v') \equiv F(u,v),\ G'(u',v') \equiv G(u,v), \\ u = u(u',v'),\ v = v(u',v') \end{array} \right. \tag{5.16}$$

The metric forms (5.11) and (5.12), satisfying to the condition (5.16), is called **isometrically identical.** Diffeomorphisms, which are **isometries,** retain the same values of arc lengths and angles between the arcs.

Note that identical isometric metric forms are also automatically isometrically equivalent, the converse is not always true.

Definition 5.3. The diffeomorphism (5.10) is called **conformal** (*angles between the arcs remain the same*), and (5.11) and (5.12) are called the **metric forms, preserving conformality**, if under the action of the diffeomorphism (5.10) $f : u = u(u',v'),\ v = v(u',v')$, the metric form (5.11)

$(ds)^2 = E(u,v)(du)^2 + 2F(u,v)\,dudv + G(u,v)(dv)^2$ differs from the metric form (5.12)

$(ds')^2 = E'(u',v')(du')^2 + 2F'(u',v')du'dv' + G'(u',v')(dv')^2$ by the positive factor $m = m(u,v) > 0$. The factor m is called a *coefficient of conformality*.

In this case, the metric form (5.12), under the action of the diffeomorphism

$$f : u = u(u',v'),\ v = v(u',v'),\ \frac{\partial(u,v)}{\partial(u',v')} \neq 0,$$

is converted to the metric form

$$(ds')^2 = E'(u',v')(du')^2 + 2F'(u',v')du'dv' + G'(u',v')(dv')^2, \qquad (5.17)$$

with the following coefficients:

$$\begin{cases} E'(u',v') \equiv m^2 E(u,v) \\ F'(u',v') \equiv m^2 F(u,v) \\ G'(u',v') \equiv m^2 G(u,v) \end{cases} \qquad (5.18)$$

what corresponds to the following identities:

$$\begin{cases} \dfrac{E'(u',v')}{E(u,v)} \equiv \dfrac{F(u',v')}{F(u,v)} \equiv \dfrac{G(u',v')}{G'(u,v)} \equiv m^2 \\ u = u(u',v'),\ v = v(u',v') \end{cases} \qquad (5.19)$$

Definition 5.4. Mapping $f : u = u(u',v'),\ v = v(u',v'),\ \dfrac{\partial(u,v)}{\partial(u',v')} \neq 0$ is called

equireal (*saves areas*), if

$$\frac{\partial(u,v)}{\partial(u',v')} \equiv \sqrt{\frac{E'G' - (F')^2}{EG - (F)^2}} \ . \qquad (5.20)$$

5.3.5. Gaussian curvature

Gaussian curvature as a measure of the **deformation** of the surface is another important notion of the internal geometry. We will not give a precise definition of this notion (for more detail, see, for example, [5]). We only note that, if we know the metric form $(ds)^2 = E(u,v)(du)^2 + 2F(u,v)dudv + G(u,v)(dv)^2$, then the Gaussian curvature $K = K(u,v)$ is calculated by the formula:

$$\begin{cases} K = \dfrac{\Delta_1 - \Delta_2}{\left(EG - F^2\right)^2} \\ \Delta_1 = \det\left(a_{ij}\right),\ \Delta_2 = \det\left(b_{ij}\right),\ i,j = 1,2,3 \end{cases} \qquad (5.21)$$

where

$$a_{11} = -\frac{1}{2}\frac{\partial^2 G}{(\partial u)^2} + \frac{\partial^2 F}{\partial u \partial v} - \frac{1}{2}\frac{\partial^2 E}{(\partial v)^2}, \ a_{12} = \frac{\partial E}{\partial u}, \ a_{13} = \frac{\partial F}{\partial u} - \frac{1}{2}\frac{\partial E}{\partial v},$$

$$a_{21} = \frac{\partial F}{\partial v} - \frac{1}{2}\frac{\partial G}{\partial u}, \ a_{22} = E, \ a_{23} = F,$$

$$a_{31} = \frac{1}{2}\frac{\partial G}{\partial u}, \ a_{32} = F, \ a_{23} = G,$$

$$b_{11} = 0, \ b_{12} = \frac{\partial E}{\partial v}, \ b_{13} = \frac{\partial G}{\partial u},$$

$$b_{21} = \frac{1}{2}\frac{\partial E}{\partial v}, \ b_{22} = E, \ b_{23} = F,$$

$$b_{31} = \frac{1}{2}\frac{\partial G}{\partial u}, \ b_{32} = F, \ b_{33} = G.$$

In our situation, we consider the metric forms, for which we have: $F = F(u,v) \equiv 0$, and then, according to this remark, we get from (5.21) the following formula for the Gaussian curvature:

$$\begin{cases} K = K(u,v) = -\frac{1}{A \bullet B}\left[\frac{\partial}{\partial u}\left(\frac{\frac{\partial B}{\partial u}}{A} \right) + \frac{\partial}{\partial v}\left(\frac{\frac{\partial A}{\partial v}}{B} \right) \right]. \\ A = \sqrt{E(u,v)}; \ B = \sqrt{G(u,v)} \end{cases} \tag{5.22}$$

5.3.6. A notion of the *distance*

Let us introduce the concept of the **distance** ρ_{12} between Lobachevski's metric form (5.8) and two-parametric metric form (5.9).

Definition 5.5. The following number:

$$\rho_{12} = R\sqrt{(\alpha-1)^2 + (\beta-1)^2} \tag{5.23}$$

is called a *distance* between Lobachevski's metric form (5.8)

$$(ds)^2 = R^2\left[(du)^2 + sh^2(u)(dv)^2 \right]$$

and the two-parametric metric form (5.9)

$$(ds')^2 = R^2\left[\alpha^2(du')^2 + \beta^2 sh^2(\alpha u')(dv')^2 \right],$$

where

$$0 < \alpha, \beta < +\infty, \; f : u = \alpha u', \; v = \beta v',$$

$$(u,v), (u',v') \in \Pi^+ : 0 < U < +\infty, -\infty < V < +\infty.$$

Because in the future, we measure the distance between the forms (5.8) and (5.9) for the same values $R > 0$, then between these forms, unless otherwise stated, it is more convenient to use the **normalized distance**:

$$\bar{\rho}_{12} = \frac{\rho_{12}}{R} = \sqrt{(\alpha - 1)^2 + (\beta - 1)^2} \; .$$

Note that for the case $\alpha = 1, \beta = 1$ we have $\rho_{12} = 0 \Leftrightarrow \bar{\rho}_{12} = 0$, $\rho_{12} > 0 \Leftrightarrow \bar{\rho}_{12} > 0$ and then the metric forms (5.8) and (5.9) have the same form, and therefore they are **isometrically identical**, here the **isometry** has the following form: $f : u = u', v = v'$.

For brevity, we will say that for the case $\bar{\rho}_{12} = 0$ the metric form (5.9) **coincides** with Lobachevski's metric form (5.8). For the case $\bar{\rho}_{12} > 0$, the metric form (5.9) **does not coincide** with Lobachevski's metric form (5.8). In this case, either both numbers α, β is not equal to 1, or one of these numbers is equal to 1 and another number is not equal to 1.

We say that the metric form (5.9) is ε-**close** to Lobachevski's metric form (5.8), if for any $\varepsilon > 0$ there is $\delta = \delta(\varepsilon) > 0$ such that for all $|\alpha - 1| < \delta, |\beta - 1| < \delta$ the inequality $0 < \bar{\rho}_{12} < \varepsilon$ exists.

Further, for the case $\bar{\rho}_{12} > 0$ we compare Lobachevski's metric form (5.8) and the two-parametric form (5.9), obtained from (5.8) under the action of the transformation (5.6), for **compatibility** or **incompatibility** of the following properties of the interior geometry: *Gaussian curvature, isometric equivalence, isometric identity, conformity, conservation of areas* .

5.3.7. Verifications

Verification to match the Gaussian curvature for the case $\bar{\rho}_{12} > 0$

1. Lobachevski's metric form (5.8):

$$(ds)^2 = R^2[(du)^2 + sh^2(u)(dv)^2] \; ,$$

$$\begin{cases} E = E(u,v) = R^2 > 0, F = F(u,v) = 0 \\ G = G(u,v) = R^2 sh^2(u) > 0 \\ A = \sqrt{E} = R > 0, B = \sqrt{G} = Rsh(u) > 0 \end{cases}$$

$$\frac{\partial B}{\partial u} = Rch(u) > 0, \quad \frac{\partial^2 B}{(\partial u)^2} = Rch(u) > 0, \quad, \quad \frac{\partial A}{\partial v} = 0.$$

Gaussian curvature of Lobachevski's metric form:

$$K = -\frac{1}{AB}\left(\frac{\partial}{\partial u}\left(\frac{\frac{\partial B}{\partial u}}{A}\right) + \frac{\partial}{\partial v}\left(\frac{\frac{\partial A}{\partial v}}{B}\right)\right) = -\frac{1}{A^2 B}\frac{\partial^2 B}{(\partial u)^2} = -\frac{1}{R^2(Rsh(u))}Rsh(u) = -\frac{1}{R^2} < 0 \tag{5.24}$$

2. Two-parametric metric form (5.9), induced by the action of the transformation (5.6) on Lobachevski's metric form (5.8).

Here the induced two-parametric metric form (5.9) has the following form:

$$(ds')^2 = R^2[\alpha^2 (du')^2 + \beta^2 (\alpha u')(du')^2],$$

$$E' = E'(u',v') = R^2\alpha^2 > 0, F' = F'(u',v') = 0,$$
$$G' = G'(u',v') = R^2\beta^2 sh^2(\alpha u') > 0$$

$$A' = \sqrt{E'} = R\alpha > 0, B' = \sqrt{G'} = R\beta sh(\alpha u') > 0,$$

$$\frac{\partial B'}{\partial u'} = R\alpha\beta ch(\alpha u'), \quad \frac{\partial^2 B'}{(\partial u')^2} = R\alpha^2\beta sh(\alpha u'), \quad \frac{\partial A'}{\partial v} = 0.$$

The Gaussian curvature of the two-parametric metric form is the following:

$$K' = -\frac{1}{A'B'}\left(\frac{\partial}{\partial u'}\left(\frac{\frac{\partial B'}{\partial u'}}{A'}\right) + \frac{\partial}{\partial v'}\left(\frac{\frac{\partial A'}{\partial v'}}{B'}\right)\right) = -\frac{1}{(A')B'}\frac{\partial^2 B'}{(\partial u')^2} =$$

$$= -\frac{1}{R^2\alpha^2(R\beta sh(\alpha u'))}R\alpha^2\beta sh(\alpha u') = -\frac{1}{R^2} < 0.$$

Conclusion. For the case $\bar{\rho}_{12} > 0$ the **Gaussian curvature** of the two-parametric metric form (5.9) for any $0 < \alpha, \beta < +\infty$ **coincides** with the **Gaussian curvature** of Lobachevski's metric form (5.8) and equal $K = -\frac{1}{R^2} < 0$.

Verification on isometric equivalence for the case $\bar{\rho}_{12} > 0$

1. Lobachevski's metric form (5.8)

$$(ds)^2 = E(u,v)(du)^2 + G(u,v)(dv)^2$$

$$E = E(u,v) = R^2 > 0, F = F(u,v) = 0$$

$$G = G(u,v) = R^2 sh^2(u) > 0$$

2. Two-parametric metric form (5.9), induced by the action of the transformation (5.6) on Lobachevski's metric form (5.8):

$$(ds')^2 = R^2[\alpha^2(du')^2 + \beta^2 sh^2(\alpha u')(dv')^2],$$

$$u = \alpha u', \; v = \beta v', \; a_1 = \frac{\partial u}{\partial u'} = \alpha, \; a_2 = \frac{\partial u}{\partial v'} = 0, b_1 = \frac{\partial v}{\partial u'} = 0, \; b_2 = \frac{\partial v}{\partial v'} = \beta,$$

$$E' = E'(u',v') = R^2\alpha^2 > 0, F' = F'(u',v') = 0, \; G' = G'(u',v') = R^2\beta^2 sh^2(\alpha u' = u) > 0$$

We verify the feasibility of the identities (5.14) on isometric equivalence:

$$\begin{pmatrix} E'(u',v') \\ F'(u',v') \\ G'(u',v') \end{pmatrix} \equiv \begin{pmatrix} a_1^2 & 2a_1b_1 & b_1^2 \\ a_1b_1 & a_1b_2+a_2b_1 & b_1b_2 \\ a_2^2 & 2a_2b_2 & b_2^2 \end{pmatrix} \begin{pmatrix} E(u,v) \\ F(u,v) \\ G(u,v) \end{pmatrix},$$

For the case $\bar{\rho}_{12} > 0$, the formula (5.14) takes the following form:

$$\begin{pmatrix} R^2\alpha^2 \\ 0 \\ R^2\beta^2 sh^2(\alpha u' = u) \end{pmatrix} \equiv \begin{pmatrix} \alpha^2 & 0 & 0 \\ 0 & \alpha\beta & 0 \\ 0 & 0 & \beta^2 \end{pmatrix} \begin{pmatrix} R^2 \\ 0 \\ R^2 sh^2(u) \end{pmatrix}$$

From here, we get the following identities:

$$R^2\alpha^2 \equiv \alpha^2 R^2 \Rightarrow \alpha \equiv \alpha, \; 0 \equiv \alpha\beta \bullet 0, \; R^2\beta^2 sh^2(\alpha u' = u) \equiv \beta^2 R^2 sh^2(u) \Rightarrow \beta \equiv \beta$$

Conclusion. For the case $\bar{\rho}_{12} > 0$ the transformations (5.6) $f : u = \alpha u'$, $v = \beta v'$ of the half-plane $\Pi^+ : 0 < U < +\infty, -\infty < V < +\infty$ are **isometric mapping,** here Lobachevski's metric form (5.8) and the two-parametric metric forms (5.9), under the action of this transformation on (5.8), are **isometrically equivalent** for any $\alpha, \beta > 0$ such that either both numbers α, β is not equal to 1, or one of these numbers equals to 1 and another number not equals to 1.

Verification on isometric identity for the case $\bar{\rho}_{12} > 0$, $\alpha \neq \beta$, $0 < \alpha, \beta < +\infty$

1. Lobachevski's metric form (5.8):

$$(ds)^2 = E(u,v)(du)^2 + G(u,v)(dv)^2$$

$$E = E(u,v) = R^2 > 0,\, F = F(u,v) = 0,$$
$$G = G(u,v) = R^2 sh^2(u) > 0$$

2.Two-parametric metric form (5.9), induced by the action of the transformation (5.6) on Lobachevski's metric form (5.8), has the following form:

$$(ds')^2 = R^2\left[\alpha^2 (du')^2 + \beta^2 sh^2(\alpha u')(dv')^2\right],$$

$$u = \alpha u',\, v = \beta v',$$

$$E' = E'(u',v') = R^2\alpha^2 > 0,\, F' = F'(u',v') = 0,$$

$$G' = G'(u',v') = R^2\beta^2 sh^2(\alpha u' = u) > 0.$$

By definition 5.2, at the **isometry** the transformation (5.6) $f : u = \alpha u'$, $v = \beta v'$, under the action on the metric form (5.6), should induce the metric form of the same type as (5.6), that is, the form (5.9) must have the form:

$$(ds')^2 = R^2(du')^2 + sh(u')(dv')^2$$

$$E' = E'(u',v') = R^2 > 0,\, F' = F'(u',v') = 0,$$
$$G' = G'(u',v') = R^2 sh^2(u' = u) > 0$$

Then, we get:

$$E' = E' = E'(u',v') = R^2\alpha^2 \equiv R^2 \Rightarrow \alpha = 1,$$

$$G' = G'(u',v') = R^2\beta^2 sh^2(\alpha u' = u) \equiv$$
$$\equiv R^2 sh^2(u' = u) \Rightarrow \beta = 1$$

For the case $\alpha = 1$, $\beta = 1$ we get $\bar{\rho}_{12} = 0$ what contradicts to the condition $\bar{\rho}_{12} > 0$.

Conclusion. For the case $\bar{\rho}_{12} > 0$ the transformation (5.6) $u = \alpha u'$, $v = \beta v'$ of the half-plane $\Pi^+ : 0 < U < +\infty, -\infty < V < +\infty$ at the condition (5.7) $0 < \alpha, \beta < +\infty$ **is not isometry**, here Lobachevski's metric form (5.8) and the two-parametric metric forms (5.9) under the action of this transformation on (5.8) **are not identical isometrically.**

1. Lobachevski's metric form (5.8):

$$(ds)^2 = E(u,v)(du)^2 + G(u,v)(dv)^2$$

$$E = E(u,v) = R^2 > 0, \, F = F(u,v) = 0,$$

$$G = G(u,v) = R^2 sh^2(u) > 0$$

2. Two-parametric metric form (5.9), induced by the action of the transformation (5.6) on Lobachevski's metric form (5.8), has the following form:

$$(ds')^2 = R^2 \left[\alpha^2 (du')^2 + \beta^2 sh^2(\alpha u')(dv')^2 \right],$$

$$u = \alpha u', \, v = \beta v',$$

$$E' = E'(u',v') = R^2 \alpha^2 > 0, \, F' = F'(u',v') = 0,$$

$$G' = G'(u',v') = R^2 \beta^2 sh^2(\alpha u' = u) > 0.$$

By virtue of the definition 5.3, in order that the mapping $f : u = \alpha u', \, v = \beta v'$ was *conformal*, the following identities should be fulfilled:

$$E'(u',v') \equiv m^2 E(u,v), \, F'(u',v') \equiv m^2 F(u,v), \, G'(u',v') \equiv m^2 G(u,v),$$

where $u = \alpha u', \, v = \beta v'$ and the coefficient of conformality $m = m(u,v) > 0$ is any function.

For our situation when $\bar{\rho}_{12} > 0$ and at the assumptions about conformality, we get:

$$E'(u',v') = R^2 \alpha^2 \equiv m^2 E(u,v) = m^2 R^2 \Rightarrow \alpha = m, \qquad (5.25)$$

$$F'(u',v') = 0 \equiv m^2 F(u,v) = m^2 \bullet 0 \Rightarrow 0 = 0, \qquad (5.26)$$

$$G'(u',v') = R^2 \beta^2 sh^2(\alpha u') \equiv m^2 E(u,v) = m^2 R^2 sh^2(u) \Rightarrow \beta \, sh(\alpha u') \equiv m \, sh(u) \qquad (5.27)$$

In virtue of the first identity (5.25), we have: $\alpha = m$, therefore the mapping $f : u = \alpha u', \, v = \beta v'$ takes the following form: $f : \, u = mu', v = \beta v'$. The second identity (5.26) does not affect on the value $m > 0$. However, from the third identity (5.27) and the condition $u > 0, m > 0$ we get:

$$\beta \bullet sh(\alpha u') \equiv m \bullet sh(u) \Rightarrow \beta \bullet sh(mu' = u) \equiv m \bullet sh(u) \Rightarrow \beta = m \qquad (5.28)$$

But then $\alpha = \beta = m > 0$, what contradicts to the condition $\bar{\rho}_{12} > 0, \alpha \neq \beta$.

Conclusions. For the case $\bar{\rho}_{12} > 0$, $\alpha \neq \beta$, the **transformations** $f : u = \alpha u'$, $v = \beta v'$ of the half-plane $\Pi^+ : 0 < U < +\infty, -\infty < V < +\infty$ for the condition $0 < \alpha, \beta < +\infty$

are not conformal, and the metric forms (5.8) and (5.9), under the action of this transformation, **are not conformal metric forms.**

For the case $\bar{\rho}_{12} > 0$, $\alpha = \beta = m \ne 1$, the transformations $f : u = mu', v = mv'$ are **conformal** with the real coefficients of conformality $m > 0, m \ne 1$. For this case the distance

$\bar{\rho}_{12} = \sqrt{2}|m-1| > 0$. For the case $\alpha = \beta = m = 1$, $\bar{\rho}_{12} = 0$.

Verification on equiareality for the case $\bar{\rho}_{12} > 0$

1. Lobachevski's metric form (5.8):

$$(ds)^2 = E(u,v)(du)^2 + G(u,v)(dv)^2$$

$$E = E(u,v) = R^2 > 0, \; F = F(u,v) = 0$$

$$G = G(u,v) = R^2 sh^2(u) > 0$$

2. Two-parametric metric form (5.9), induced by the action of the transformation (5.6) on Lobachevski's metric form (5.8), has the following form:

$$(ds)^2 = R^2[\alpha^2(du')^2 + \beta^2 sh(\alpha u')(dv')^2],$$

$$f : u = \alpha u', \; v = \beta v',$$

$$E' = E'(u',v') = R^2\alpha^2 > 0, \; F' = F'(u',v') = 0,$$

$$G' = G'(u',v') = R^2\beta^2 sh^2(\alpha u' = u) > 0$$

According to the definition 5.4, for the case of **equiareality,** the diffeomorphism $f : u = u(u',v'), \; v = v(u',v')$, $\dfrac{\partial(u,v)}{\partial(u',v')} \ne 0$ must satisfy to the condition (5.20)

In our situation, for the case $\bar{\rho}_{12} > 0$ for the diffeomorphism (5.6) $f : u = \alpha u'$, $v = \beta v'$ of the half-plane $\Pi^+ : 0 < U < +\infty, -\infty < V < +\infty$ at the condition (5.7) $0 < \alpha, \beta < +\infty$ we get that $\dfrac{\partial u}{\partial u'} = \alpha > 0$, $\dfrac{\partial u}{\partial v'} = 0$, $\dfrac{\partial v}{\partial u'} = 0$, $\dfrac{\partial v}{\partial v'} = \beta > 0$, whence we have:

$$\frac{\partial(u,v)}{\partial(u',v')} = \alpha\beta > 0 \tag{5.29}$$

On the other hand, we have:

$$\sqrt{\frac{E'G'-(F')^2}{EG-(F)^2}} = \sqrt{\frac{R^4\alpha^2\beta^2 sh^2(\alpha u' = u)}{R^4 sh^2(u)}} = \alpha\beta \tag{5.30}$$

169

Comparing (5.28) and (5.9), we get that (5.20) holds for all $0 < \alpha, \beta < +\infty$, in which $\bar{\rho}_{12} = \sqrt{(\alpha-1)^2 + (\beta-1)^2}$, that is, the two-parametric form (5.9) does **not coincide** with Lobachevski's metric form (5.8)

Conclusions. For the case $\bar{\rho}_{12} > 0$, the **transformations** (5.6) $f : u = \alpha u'$, $v = \beta v'$ of the half-plane Π^+: $0 < U < +\infty, -\infty < V < +\infty$ for the case (5.7) $0 < \alpha, \beta < +\infty$ are **equiareal**.

Conclusions on the intrinsic properties of metric forms

Individual properties (the *conservation* of the **Gaussian curvature**, **isometric equivalence, equiareality**, but at the same time the *non-conservation* of **isometric identity** and **conformity**), associated with each of the infinite set of the metric form (5.9) with the same Gaussian curvature $K = -\dfrac{1}{R^2} < 0$, appears not only at the comparison of these metric forms, by using the transformations (5.6), with Lobachevski's classical metric form (5.8), but also at the comparison (with certain limitations) of the metric forms (5.9) among themselves, by using the transformations similar to (5.6). It is important to emphasize that for the case $\alpha, \beta \to 1$ **all the metric forms** (5.9) **are close to the Lobachevski's metric form** (5.8).

5.4. Lobachevski's metric (λ,μ)-forms)

Below, in order not to overload the text, we will consider the metric forms having **one and the same** Gaussian curvature $K = -\dfrac{1}{R^2} = -1 \Leftrightarrow R = 1$ what does not affect on the generality of the results. Then, Lobachevski's metric form (5.8) and two-parametric metric form (5.9) take the following forms:

$$(ds)^2 = (du)^2 + sh^2(u)(dv)^2 \tag{5.31}$$

$$(ds')^2 = \alpha^2(du')^2 + \beta^2 sh^2(\alpha u')(dv')^2 \tag{5.32}$$

Because $R = 1$, then for this situation the distance ρ_{12} of the kind (5.23) between the metric forms (5.31) и (5.32) coincides with the normalized distance $\bar{\rho}_{12} = \dfrac{\rho_{12}}{R} = \sqrt{(\alpha-1)^2 + (\beta-1)^2}$:

As for the diffeomorphism f of the half-plane Π^+: $0<U<+\infty, -\infty<V<+\infty$, under the action of which the metric form (5.31) is converted into the metric form (5.32), it has the previous form (5.6) $f: u = \alpha u', v = \beta v'$, where α, β are some real numbers, for the conditions (5.7) $0<\alpha, \beta<+\infty$.

Let us represent the diffeomorphism (5.6) and the metric form (5.32) in the terms of the **"metallic proportions"** and **"hyperbolic Fibonacci functions,"** discussed in previous chapters, because these objects are of great theoretical and practical importance.

Recall that the **"metallic proportions"** are called the real numbers of the form $\Phi_\lambda = \dfrac{\lambda + \sqrt{4+\lambda^2}}{2}$, $\lambda \neq 0$, while for the first four **positive integers** $\lambda = 1, 2, 3, 4$ we use the following names: $\Phi_1 = \dfrac{1+\sqrt{5}}{2}$ is the **golden proportion** (the **golden ratio**), $\Phi_2 = 1+\sqrt{2}$ is the **silver proportion**, $\Phi_3 = \dfrac{3+\sqrt{13}}{2}$ is the **bronze proportion**, $\Phi_4 = 2+\sqrt{5}$ is the **copper** proportion.

The following functions are called the **hyperbolic Fibonacci λ-sine** $sF_\lambda(x)$ and the **hyperbolic Fibonacci λ-cosine** $cF_\lambda(x)$, respectively:

$$\begin{cases} sF_\lambda(x) = \dfrac{\Phi_\lambda^x - \Phi_\lambda^{-x}}{\sqrt{4+\lambda^2}} \\[3mm] cF_\lambda(x) = \dfrac{\Phi_\lambda^x + \Phi_\lambda^{-x}}{\sqrt{4+\lambda^2}} \end{cases} \tag{5.33}$$

Since $\Phi_\lambda^x = e^{x \ln(\Phi_\lambda)}$, we get the following relationship between the classical hyperbolic functions $sh(x)$, $ch(x)$ and Fibonacci λ–functions $sF_\lambda(x)$, $cF_\lambda(x)$:

$$\begin{cases} sF_\lambda(x) = \dfrac{2}{\sqrt{4+\lambda^2}} sh[(\ln \Phi_\lambda)x] \\[3mm] cF_\lambda(x) = \dfrac{2}{\sqrt{4+\lambda^2}} ch[(\ln \Phi_\lambda)x] \end{cases} \tag{5.34}$$

$$\begin{cases} sh(x) = \dfrac{\sqrt{4+\lambda^2}}{2} sF_\lambda\left(\dfrac{x}{\ln \Phi_\lambda}\right) \\[3mm] ch(x) = \dfrac{\sqrt{4+\lambda^2}}{2} cF_\lambda\left(\dfrac{x}{\ln \Phi_\lambda}\right) \end{cases} \tag{5.35}$$

Such formulas are useful for differentiation and integration and derivation of other relations. Hence, for instance, we get:

$$\frac{d[sF_\lambda(x)]}{dx} = (\ln \Phi_\lambda)cF_\lambda(x)$$

$$\frac{d[cF_\lambda(x)]}{dx} = (\ln \Phi_\lambda)sF_\lambda(x)$$

$$\int sF_\lambda(x)dx = \frac{cF_\lambda(x)}{\ln \Phi_\lambda} + const$$

$$\int cF_\lambda(x)dx = \frac{sF_\lambda(x)}{\ln \Phi_\lambda} + const$$

$$\left[cF_\lambda(x)\right]^2 - \left[sF_\lambda(x)\right]^2 = \frac{4}{4+\lambda^2}$$

In the terms of the **"metallic proportions"** and **"hyperbolic Fibonacci functions,"** the diffeomorphism (5.6) and the metric form (5.32) have the following forms:

$$f : u = \ln(\Phi_\lambda)u' > 0, \ v = \ln(\Phi_\mu)v' \tag{5.36}$$

$$(ds')^2 = \ln^2(\Phi_\lambda)(du')^2 + \ln^2(\Phi_\mu)\frac{4+\lambda^2}{4}\left[sF_\lambda(u')\right]^2 (dv')^2 \ , \tag{5.37}$$

where the old parameters $\alpha, \beta \ (0 < \alpha, \beta < +\infty))$ are connected with the new parameters $\lambda, \mu \ (0 < \lambda, \mu < +\infty)$ as follows:

$$\alpha = \ln(\Phi_\lambda) = \ln\left(\frac{\lambda + \sqrt{4+\lambda^2}}{2}\right), \ \ \beta = \ln(\Phi_\mu) = \ln\left(\frac{\mu + \sqrt{4+\mu^2}}{2}\right) . \tag{5.38}$$

The graph of the function $\alpha = \alpha(\lambda) = \ln(\Phi_\lambda)$ is showed on Fig.5.4. The graph of the function $\beta = \beta(\lambda) = \ln(\Phi_\lambda)$ has a similar form.

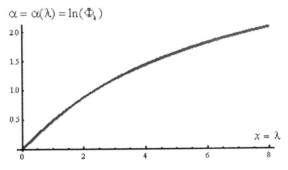

Figure 5.4. The graph of the function $\alpha = \alpha(\lambda) = \ln(\Phi_\lambda)$

The two-parametric metric forms (5.37) are called **Lobachevski's metric (λ,μ)-forms.** In these terms, the distance ρ_{12} between the **Lobachevski's metric (λ,μ)-forms** (5.37) and **Lobachevski's classic metric form** (5.31) takes the following form:

$$\rho_{12} = \sqrt{(\alpha-1)^2 + (\beta-1)^2} = \sqrt{\left[\ln(\Phi_\lambda)-1\right]^2 + \left[\ln(\Phi_\mu)-1\right]^2}.$$

Note that $\rho_{12} = 0 \Leftrightarrow \lambda = \mu = 0 \Leftrightarrow \lambda = \mu = 2sh(1) \approx 2.3504$, that is, when $\lambda = \mu \to 2sh(1)$, the metric forms (5.37) are arbitrarily close to Lobachevski's classic metric form (5.31). In particular, if the parameters (λ,μ) take only the *integer values* 1,2,3, ..., we get a **countable set** of the metric forms (5.37) of the same Gaussian curvature $K = -1$. For the case $\lambda = \mu$, the metric forms (5.37) are called the **"pure Lobachevski's metric** $(\lambda,\mu)-$**forms**, and for the case $\lambda \neq \mu$ they are called the the **"mixed Lobachevski's metric** $(\lambda,\mu)-$**forms.**

The pure Lobachevski's metric $(\lambda,\mu)-$forms:

1. The "golden" $(\lambda,\mu)-$form for the case $\lambda = \mu = 1$, $\rho_{12} \approx 0.7336$

2. The "silver" $(\lambda,\mu)-$form for the case $\lambda = \mu = 2$, $\rho_{12} \approx 0.1677$

3. The "bronze" $(\lambda,\mu)-$form for the case $\lambda = \mu = 3$, $\rho_{12} \approx 0.2754$

4. The "copper" $(\lambda,\mu)-$form for the case $\lambda = \mu = 4$, $\rho_{12} \approx 0.6273$

The mixed Lobachevski's metric $(\lambda,\mu)-$forms:

1. The "golden-silver" $(\lambda,\mu)-$form for the case $\lambda = 1, \mu = 2$, $\rho_{12} \approx 0.5321$

2. The "golden-bronze" $(\lambda,\mu)-$form for the case $\lambda = 1, \mu = 3$, $\rho_{12} \approx 0.5541$

3. The "golden-copper" $(\lambda,\mu)-$form for the case $\lambda = 1, \mu = 4$, $\rho_{12} \approx 0.6826$

4. The "copper-golden" $(\lambda,\mu)-$form for the case $\lambda = 4, \mu = 1$, $\rho_{12} \approx 0.6826$

5. The "copper-silver" $(\lambda,\mu)-$form for the case $\lambda = 4, \mu = 2$, $\rho_{12} \approx 0.4592$

6. The "copper-golden" $(\lambda,\mu)-$form for the case $\lambda = 4, \mu = 3$, $\rho_{12} \approx 0.4845$

5.5. Geodesic lines for the metric (λ,μ)-forms of Lobachevski's plane and other geometric objects

5.5.1. Preliminary information

Let us consider the half-plane $\Pi^+ : 0 < U < +\infty, -\infty < V < +\infty$, where Lobachevski's metric form (5.31) has the following form: $(ds)^2 = (du)^2 + sh^2(u)(dv)^2$. Note that we consider the situation for the case of the Gaussian curvature $K = -\dfrac{1}{R^2} = -1 \Leftrightarrow R = 1$.

To find the geodesic lines, we consider the surface:

$$M^2: Z^2 - X^2 - Y^2 = 1, \ Z \geq 1, \tag{5.39}$$

which is the upper half of the two-sheeted hyperboloid (pseudo sphere of the radius $R=1$), embedded in the three-dimensional Euclidean space (X, Y, Z), having Minkowski's metric $(dl)^2 = (dZ)^2 - (dX)^2 - (dY)^2$ with the following *parameterization* of the surface (5.39) in the form:

$$M^2 : X = sh(u)\cos(v), Y = sh(u)\sin(v), Z = ch(u) \tag{5.40}$$

Geodesic lines L on M^2 can be found in the form of the intersection $L = M^2 \cap \Pi$, where the Π-surfaces pass through the coordinate origin O $(0,0,0)$ of the space (X, Y, Z)

$$\Pi : \ aX + bY + cZ = 0. \tag{5.41}$$

Then the intersection $L = M^2 \cap \Pi$ satisfies to the system of equations:

$$Z^2 - X^2 - Y^2 = 1, \ Z \geq 1, \ aX + bY + cZ = 0. \tag{5.42}$$

Let us consider two cases: 1) $c=0$; 2) $c \neq 0$.

For the case 1) $c=0$, we get $\Pi : aX + bY = 0 \Rightarrow a\cos(v) + b\sin(v) = 0$, that is, the geodesic lines are half-lines

$$S : u > 0, \ v = const. \tag{5.43}$$

For the case 2) $c \neq 0$, we get: Π: $Z = AX + BY$, $A = -\dfrac{a}{c}$, $B = -\dfrac{b}{c}$

Hence, we obtain the relations: $Z = ch(u) = sh(u)[A\cos(v) + B\sin(v)]$, it follows from here that the geodesic lines have the following arcs:

$$S : A\cos(v) + B\sin(v) = cth(u), u > 0, -\infty < v < +\infty .$$

For other equivalent record, the last relation has the following form:

$$\begin{cases} S : u = u(v) = \dfrac{1}{2}\ln\left(\dfrac{A\cos(v)+B\sin(v)+1}{A\cos(v)+B\sin(v)-1}\right), \\ u > 0, -\infty < v < +\infty \end{cases} \qquad (5.44)$$

where $|A\cos(v)+B\sin(v)| > 1$.

The surface (5.39) M^2: $Z^2 - X^2 - Y^2 = 1$, $Z \geq 1$, in addition to the parameterization (5.40), also allows the parameterization of the following form:

$$M^2 : X = sh(\alpha u')\cos(\beta v'),\ Y = sh(\alpha u')\sin(\beta v'),\ Z = ch(\alpha u'). \qquad (5.45)$$

But then the diffeomorphism $f : \Pi^+ \to \Pi^+$ of the half-surface
$\Pi^+ : 0 < U < +\infty, -\infty < V < +\infty$ of the kind $f : u = \alpha u',\ v = \beta v'\ (0 < \alpha, \beta < +\infty)$ converts the geodesic lines S of the form (5.43) and (5.44) of Lobachevski's metric form
$(ds)^2 = (du)^2 + sh^2(u)(dv)^2$ into the geodesic lines $S' = f(S)$ of the metric form
$(ds')^2 = \alpha^2 (du')^2 + \beta^2 sh^2(\alpha u')(dv')^2$.

In the coordinates (u',v') the **geodesic lines** $S' = f(S)$ have the following form:

1). $u' > 0$, $v' = const$, if in (5.42) $c = 0$;

2). $A\cos(\beta v') + B\sin(\beta v') = cth(\alpha u'),\ u' > 0, -\infty < v' < +\infty$,

$A = -\dfrac{a}{c}$, $B = -\dfrac{b}{c}, c \neq 0$.

In other equivalent record the last relation can be rewritten as follows:

$$\begin{cases} u' = u'(v') = \dfrac{1}{2\alpha}\ln\left(\dfrac{A\cos(\beta v')+B\sin(\beta v')+1}{A\cos(\beta v')+B\sin(\beta v')-1}\right), \\ u' > 0, -\infty < v' < +\infty \end{cases} \qquad (5.46)$$

where $|A\cos(\beta v') + B\sin(\beta v')| > 1$.

Therefore, the diffeomorphism $f : u = \alpha u',\ v = \beta v'\ (0 < \alpha, \beta < +\infty)$ is **geodesic mapping**.

5.5.2. Presentation of the geodesic lines in the terms of the "metallic proportions" and "Fibonacci hyperbolic geometry"

After replacing $\alpha = \ln(\Phi_\lambda), \beta = \ln(\Phi_\mu)(0 < \lambda, \mu < +\infty)$ for the metric (λ, μ)-form (5.37), the **geodesic lines** $S' = f(S)$ in the coordinates (u', v') have the following form:

1). $u' > 0$, $v' = const$, if in (5.42) $c = 0$;

2). $A\cos[\ln(\Phi_\mu)v'] + B\sin[\ln(\Phi_\mu)v'] = \dfrac{\Phi_\lambda^{u'} + \Phi_\lambda^{-u'}}{\Phi_\lambda^{u'} - \Phi_\lambda^{-u'}}$, $u' > 0, -\infty < v' < +\infty$

$A = -\dfrac{a}{c}$, $B = -\dfrac{b}{c}$, if in (5.42) $c \neq 0$.

In other equivalent record the last relation can be rewritten as follows:

$$\begin{cases} u' = u'(v') = \dfrac{1}{2\ln(\Phi_\lambda)}\ln\left(\dfrac{A\cos[\ln(\Phi_\mu)v'] + B\sin[\ln(\Phi_\mu)v'] + 1}{A\cos[\ln(\Phi_\mu)v'] + B\sin[\ln(\Phi_\mu)v'] - 1}\right), \\ u' > 0, -\infty < v' < +\infty \end{cases} \qquad (5.47)$$

where $\left|A\cos[\ln(\Phi_\mu)v'] + B\sin[(\Phi_\mu)v']\right| > 1$.

5.6. Poincare's model of Lobachevski's plane on the unit disc and connection between Poincare's model of Lobachevski's plane and the (λ, μ)-models of Lobachevski's plane

5.6.1. The angles between the curves for the metric forms for general case

For the completeness of presentation, we recall the well-known formulas of the internal geometry for each fixed metrical form:

$$(ds)^2 = E(u,v)(du)^2 + 2F(u,v)dudv + G(u,v)(dv)^2, \qquad (5.48)$$

$$E = E(u,v) > 0, G = G(u,v) > 0, EG - F^2 > 0.$$

Suppose that $L_1 : u = u_1(t)$, $v = v_1(t)$ and $L_2 : u = u_2(t)$, $v = v_2(t)$ are two smooth arcs, which intersect at the value of the parameter $t = 0$ in the point (u_0, v_0). Then, the angle θ between these arcs in the point (u_0, v_0) is defined as follows:

$$\sin(\theta) = \sqrt{E_0 G_0 - (F_0)^2} \, \frac{A_1}{\sqrt{B_1}\sqrt{B_2}}, \ \cos(\theta) = \frac{A_2}{\sqrt{B_1}\sqrt{B_2}},$$

where

$$E_0 = E(u_0, v_0), \ F_0 = F(u_0, v_0), \ G_0 = G(u_0, v_0),$$

$$A_1 = u'_1(0)v'_2(0) - v'_1(0)u'_2(0),$$

$$A_2 = E_0 u'_1(0)u'_2(0) + F_0\left[u'_1(0)v'_2(0) + v'_1(0)u'_2(0)\right] + G_0 v'_1(0)v'_2(0),$$

$$B_1 = E_0\left[u'_1(0)\right]^2 + 2F_0\left[u'_1(0)v'_1(0)\right] + G_0\left[v'_1(0)\right]^2,$$

$$B_2 = E_0\left[u'_2(0)\right]^2 + 2F_0\left[u'_2(0)v'_2(0)\right] + G_0\left[v'_2(0)\right]^2$$

where $(.)' = \dfrac{d(.)}{dt}$ is the operator of the derivative taking, $dudv$ is the inner (scalar) product of the differentials.

Suppose $L : u=u(t), v=v(t)$ is the arc, bounded by the points of the curve, corresponding to the parameter values t_1 и t_2. Then, the length s of the arc is calculated by the formula:

$$s = \int_{t_1}^{t_2}\left[\sqrt{E(t)\left(\frac{du}{dt}\right)^2 + 2F(t)\frac{du}{dt}\frac{dv}{dt} + G(t)\left(\frac{dv}{dt}\right)^2}\right]dt,$$

where

$$E(t) = E(u(t), v(t)), \ F(t) = F(u(t), v(t)), \ G(t) = G(u(t), v(t)).$$

The surface S of the area D is calculated by the formula:

$$S = \iint_D \sqrt{EG - F^2}\, du \wedge dv,$$

where $du \wedge dv$ is external (vectorial) product of the differentials.

Recall the notion of geodesic line, which is a natural generalization of straight line on the plane. Geodesic lines $\underline{L}$ have the following extreme property: every point $(u_0, v_0) \in L$ has such neighbourhood Ω, where for any two points (u_1, v_1), $(u_2, v_2) \in L$ the arc of the curve L with ends in these points has the shortest length in comparison with any other arcs having the points (u_1, v_1), (u_2, v_2) as their endpoints.

From the standpoint of the internal geometry, the geodesic lines are such curves, along which the geodesic curvature $k_g = 0$.

Suppose that the curve L is given in parametric form:

$$L : u=u(t), \; v=v(t).$$

Then, the geodesic curvature k_g of the curve L is called the function of the following form:

$$k_g = \frac{\left(\sqrt{EG-F^2}\right)\left(\dfrac{d^2u}{dt^2}\dfrac{dv}{dt} - \dfrac{du}{dt}\dfrac{d^2v}{dt^2} + A\dfrac{dv}{dt} - B\dfrac{du}{dt}\right)}{\sqrt{\left[E\left(\dfrac{du}{dt}\right)^2 + 2F\dfrac{du}{dt}\dfrac{dv}{dt} + G\left(\dfrac{dv}{dt}\right)^2\right]^3}}$$

where

$$A = \Gamma^1_{11}\left(\frac{du}{dt}\right)^2 + 2\Gamma^1_{12}\frac{du}{dt}\frac{dv}{dt} + \Gamma^1_{22}\left(\frac{dv}{dt}\right)^2$$

$$B = \Gamma^2_{11}\left(\frac{du}{dt}\right)^2 + 2\Gamma^2_{12}\frac{du}{dt}\frac{dv}{dt} + \Gamma^2_{22}\left(\frac{dv}{dt}\right)^2$$

$$E(t)=E(u(t),v(t)), \; F(t)=F(u(t),v(t)), \; G(t)=G(u(t),v(t)).$$

The symbols Γ^k_{ij} are called *Christoffel coefficients*; they are the following forms:

$$\Gamma^1_{11} = \frac{a_1G-b_1F}{EG-F^2}, \; \Gamma^2_{11} = \frac{b_1E-a_1F}{EG-F^2},$$

$$\Gamma^1_{12} = \frac{a_2G-b_2F}{EG-F^2}, \; \Gamma^2_{12} = \frac{b_2E-a_2F}{EG-F^2},$$

$$\Gamma^1_{22} = \frac{a_3G-b_3F}{EG-F^2}, \; \Gamma^2_{22} = \frac{b_3E-a_3F}{EG-F^2},$$

where

$$a_1 = \frac{1}{2}\frac{\partial E}{\partial u}, \; b_1 = \frac{\partial F}{\partial u} - \frac{1}{2}\frac{\partial E}{\partial v},$$

$$a_2 = \frac{1}{2}\frac{\partial E}{\partial v}, \; b_2 = \frac{\partial G}{\partial u}$$

$$a_3 = \frac{\partial F}{\partial v} - \frac{1}{2}\frac{\partial G}{\partial u}, \; b_3 = \frac{1}{2}\frac{\partial G}{\partial v}$$

Because along the geodesic lines $L : u=u(t), \; v=v(t)$ the geodesic curvature $k_g=0$, then the equation of the geodesic lines $L : u=u(t), \; v=v(t)$ satisfies the differential equation:

$$\frac{d^2u}{dt^2}\frac{dv}{dt} - \frac{d^2v}{dt^2}\frac{du}{dt} + A\frac{dv}{dt} - B\frac{du}{dt} = 0 .$$

We also note that, according to Gauss-Bonnet theorem in the case of constant Gaussian curvature K the sum of the angles $\alpha + \beta + \gamma$ of any geodesic triangle, having the area $\Delta > 0$, satisfies to the relation: $\alpha + \beta + \gamma = \pi + K\Delta$. Hence, we obtain the well-known facts of non-Euclidean geometry:

$$\alpha + \beta + \gamma < \pi, \text{ if } K < 0,$$

$$\alpha + \beta + \gamma = \pi, \text{ if } K = 0,$$

$$\alpha + \beta + \gamma > \pi, \text{ if } K > 0.$$

Next, we turn to more detailed examination of the various interpretations of Lobachevski's geometry and its perturbations.

The geodesic lines for Lobachevski's metric form (5.31), as well as geodesic lines for Lobachevski's metric (λ, μ)-forms, which are on the distance $\rho_{12} = \sqrt{\left[\ln(\Phi_\lambda) - 1\right]^2 + \left[\ln(\Phi_\mu) - 1\right]^2} > 0$ from the metric form (5.31), are a special case of such smooth curves on the half-plane $0 < U < +\infty, -\infty < V < +\infty$; while for Lobachevski's metric form (5.31), and for Lobachevski's metric (λ, μ)-forms (5.37) the Gaussian curvature $K = -1$.

As we indicate above, for the case $\bar{\rho}_{12} > 0$ the diffeomorphisms $f : u = \ln(\Phi_\lambda)u'$, $v = \ln(\Phi_\lambda)v'$, which convert the metric form (5.31) into the metric forms (5.37), **are not conformal**, that is, not preserve angles between the relevant curves at the transition from Lobachevski's metric form (5.31) to Lobachevski's metric (λ, μ)-forms (5.37), under the action of diffeomorphisms f.

The authors did not set out the goal to derive the relationships between all other possible geometric and differential objects, induced by Lobachevski's metric (λ, μ)-forms of the Gaussian curvative $K = -1$ or the fixed Gaussian curvative $K < 0$ on the half-plane $(0 < U < +\infty, -\infty < V < +\infty)$, which are on the distance $\bar{\rho}_{12} > 0$ from Lobachevski's metric form (5.31) at the changes of the parameters $\lambda, \mu > 0$ and coincide with Lobachevski's metric form for the case $\lambda = 1, \mu = 1$. This problem is the subject of a special study.

In connection with this problem, we consider only the model of Lobachevski's plane on the disk $D^2 : x^2 + y^2 < 1$, proposed in 1882 by the great

French mathematician, physicist and astronomer **Henri Poincare** (1854-1912). We show its relationship to Lobachevski's metric (λ,μ) – forms.

Figure 5.5. Henri Poincaré (1854-1912)

Let us remind the basic facts of Lobachevski's geometry for Poincare's realization on the disc $D^2 x^2 + y^2 < 1$. The information is taken from [5].

Recall that for $R=1$ $K = -\dfrac{1}{R^2} = -1$, for this case the surface (5.2) has the form: $M^2 : Z^2 - X^2 - Y^2 = 1$, $Z \geq 1$ and, according to [5], also it allows a parameterization of the kind:

$$X = \frac{2x}{1-x^2-y^2}, \ Y = \frac{2y}{1-x^2-y^2}, \ Z = \frac{1+x^2+y^2}{1-x^2-y^2}, \tag{5.49}$$

where $(x,y) \in D^2 : x^2 + y^2 < 1$. For the variables (x, y), Poincare's metric form of the Gaussian curvature $K = -1$ is given by:

$$(ds)^2 = \frac{4\left[(dx)^2 + (dy)^2\right]}{\left(1-x^2-y^2\right)^2}, \tag{5.50}$$

where $(ds)^2 = -(dl)^2$, $(dl)^2 = (dZ)^2 - (dX)^2 - (dY)^2$, dl is an arc element in the space (X, Y, Z) for the surface M^2: $Z^2 - X^2 - Y^2 = 1$, $Z \geq 1$, ds is an arc element in the disc $D^2 : x^2 + y^2 < 1$.

Since, according to (5.45), the surface (5.39) M^2: $Z^2 - X^2 - Y^2 = 1$, $Z \geq 1$ also admits a parameterization of the kind M^2:

$X = sh(\alpha u')\cos(\beta v')$, $Y = sh(\alpha u')\sin(\beta v')$, $Z = ch(\alpha u')$, where $(u',v') \in \Pi^+$:

$(0 < U < +\infty, -\infty < V < +\infty)$, $\alpha, \beta > 0$, then (u', v') and (x, y) are connected by relationships:

$$\begin{cases} X = \dfrac{2x}{1 - x^2 - y^2} = sh(\alpha u')\cos(\beta v'), \\[3mm] Y = \dfrac{2y}{1 - x^2 - y^2} = sh(\alpha u')\sin(\beta v'), \\[3mm] Z = \dfrac{1 + x^2 + y^2}{1 - x^2 - y^2} = ch(\alpha u'). \end{cases} \qquad (5.51)$$

Hence, we obtain the following recalculation $(x, y) \in D^2$ into $(u', v') \in \Pi^+$:

$$\begin{cases} sh(\alpha u') = \dfrac{2\sqrt{x^2 + y^2}}{1 - \left(x^2 + y^2\right)} \\[3mm] ch(\alpha u') = \dfrac{1 + \left(x^2 + y^2\right)}{1 - \left(x^2 + y^2\right)} \end{cases}, \qquad (5.52)$$

$$\begin{cases} \sin(\beta v') = \dfrac{y}{\sqrt{x^2 + y^2}} \\[3mm] \cos(\beta v') = \dfrac{x}{\sqrt{x^2 + y^2}} \end{cases}. \qquad (5.53)$$

The inverse recalculation $(u', v') \in \Pi^+$ into $(x, y) \in D^2$ gives the following result:

$$\begin{cases} x = x(u', v') = \sqrt{\dfrac{-1 + ch(\alpha u')}{1 + ch(\alpha u')}} \cos(\beta v') \\[3mm] y = y(u', v') = \sqrt{\dfrac{-1 + ch(\alpha u')}{1 + ch(\alpha u')}} \sin(\beta v') \end{cases}. \qquad (5.54)$$

Then, we obtain the following correspondences:

$$(\ 0 < x^2 + y^2 < 1) \Leftrightarrow (u' > 0, -\infty < v' < +\infty, \ 0 < \alpha, \beta < +\infty),$$

$$\begin{cases} (x = 0, y = 0) \bigcup (x^2 + y^2 = 1) \Leftrightarrow \\ \Leftrightarrow (u' = 0, -\infty < v' < +\infty) \bigcup \\ \bigcup (u' = +\infty, -\infty < v' < +\infty) \end{cases}.$$

Recall that $0 \le x^2 + y^2 < 1$ is the domain of Poincare's metric form (5.50), for which the *absolute* is $x^2 + y^2 = 1$, and $u' > 0, -\infty < v' < +\infty$ is the domain of the definition of the two-parametric forms (5.32), for which the *absolute* is the union:

$$(u' = 0, -\infty < v' < +\infty) \cup (u' = +\infty, -\infty < v' < +\infty) \text{ (see the section 5.3.2).}$$

It follows from this that the comparison by using the transformations (5.52)-(5.54) of Poincare's metric form (5.50), defined on the disc $0 \le x^2 + y^2 < 1$, with the two-parametric forms (5.32), defined on the half-plane $u' > 0, -\infty < v' < +\infty$ ($0 < \alpha, \beta < +\infty$), is possible only for the condition, when Poincare's metric form (5.50) is considered in the domain $0 \le x^2 + y^2 < 1$.

Then, under the action of the diffeomorphism (5.54), Poincare's metric form (5.50), which is considered for this case in the ring $0 < x^2 + y^2 < 1$, is converted for each fixed values of the parameters $0 < \alpha, \beta < +\infty$ into the two-parametric form (5.32) by retaining the element of the arc length, that is,

$$(ds)^2 = \frac{4\left[(dx)^2 + (dy)^2\right]}{\left(1 - x^2 - y^2\right)^2} = \alpha^2 (du')^2 + \beta^2 sh(\alpha u')(dv')^2 = (ds')^2. \tag{5.55}$$

According to the **Definition 5.1**, the transformation (5.54) in this case is **isometric mapping** and the metric forms (5.50) and (5.32) are **isometrically equivalent**.

However, since for the variables (x, y) and (u', v') the metric forms (5.50) and (5.32) are different, then the metric forms (5.50) and (5.32) are not isometrically identical, and the transformation (5.54) is not an isometry (see the **Definition 5.2**)

Let $A_1(x_1, y_1)$ and $A_2(x_2, y_2)$ be arbitrary points of Lobachevski's plane, which is realized in the form of a circle $D^2 : 0 \le x^2 + y^2 < 1$ with the metrics (5.50).

Further we use complex numbers. We designate the point $A(x, y)$ by $z = x + iy$, where $i = \sqrt{-1}$ is imaginary unit. The module of the complex number z is equal to $|z| = \sqrt{x^2 + y^2}$. Let $\bar{z} = x - iy$ be a complex number conjugate to the complex number $z = x + iy$.

For this case the points $A_1(x_1, y_1)$ and $A_2(x_2, y_2)$ in complex notation can be represented as follows: $z_1 = x_1 + iy_1$, $z_2 = x_2 + iy_2$. It is well-known that the **distance** $\rho(A_1, A_2)$ between two points $A_1(x_1, y_1)$ and $A_2(x_2, y_2)$ in complex notation has the following form:

$$\rho(A_1, A_2) = \ln \left(\frac{1 + \left| \frac{z_1 - z_2}{z_1 - \overline{z_2}} \right|}{1 - \left| \frac{z_1 - z_2}{z_1 - \overline{z_2}} \right|} \right) \tag{5.56}$$

For real variables $0 \le x^2 + y^2 < 1$, the formula (5.56) has the following form:

$$ch[\rho(A_1, A_2)] = 1 + 2 \frac{(x_1 - x_2)^2 + (y_1 - y_2)^2}{(1 - x_1^2 - y_1^2)(1 - x_2^2 - y_2^2)}.$$

This implies that if in the area $0 \le x^2 + y^2 < 1$ we fix the point (x_1, y_1), and $x_2^2 + y_2^2 \to 1$ "from within" the domain $0 \le x^2 + y^2 < 1$, then $ch(\rho) \to +\infty$ and therefore at approaching to the *absolute*, the distance $\rho \to +\infty$, where ρ is induced by Poincare's metric form (5.50).

In complex notation the metrics (5.50) has the following form:

$$(ds)^2 = \frac{4}{\left(1 - |z|^2\right)^2} dz d\overline{z}, \ |z| < 1. \tag{5.57}$$

The *movement* for the metrics (5.50) of Lobachevski's plane can be written as follows:

$$z' = f(z) = \frac{Az + \overline{B}}{Bz + \overline{A}}, \ |A|^2 - |B|^2 = 1, \tag{5.58}$$

where $z = x + iy$ and $z' = x' + iy'$.

Note that at the movements (5.58) the metric form is saved, that is, if we consider the transformation

$$z = f^{-1}(z') = \frac{-\overline{A}z' + \overline{B}}{Bz' - A}, \ |A|^2 - |B|^2 = 1, \tag{5.59}$$

inverse with respect to the transformation (5.58), and then we substitute (5.59) into (5.57), we get the metric form

$$(ds')^2 = \frac{4}{\left(1 - |z'|^2\right)^2} dz' d\overline{z}', |z'| < 1, \text{ where } ds = ds'. \tag{5.60}$$

According to the **Definition 5.2**, the transformation (5.59) is an **isometry** and the metric form (5.57) and (5.60) are **isometrically identical**. Since some **isometry** is also a **conformal transformation** (see the definitions 2 and 3), the transformation (5.59) preserves not only the *elements of the arc lengths*, but the *angles between the arcs.* Therefore, the metric forms (5.57) and (5.60) are also **conformal metric forms,** having the same conformal forms. The transformation (5.58) has similar properties.

Consider further the form of the geodesic lines in the coordinates (x, y) for the metric form (5.50) and compare them with the geodesic lines in the coordinates (u', v') for the metric form (5.32).

According to (5.42), the geodesic lines L on M^2 with $K = -1$ satisfy to the conditions:

$$M^2 : Z^2 - X^2 - Y^2 = 1, Z \geq 1, \Pi : aX + bY + cZ = 0.$$

According to (5.49), the surface $M^2 : Z^2 - X^2 - Y^2 = 1, Z \geq 1$ admits a parameterization:

$$\begin{cases} M^2 : X = \dfrac{2x}{1-x^2-y^2}, \\ Y = \dfrac{2y}{1-x^2-y^2}, \\ Z = \dfrac{1+x^2+y^2}{1-x^2-y^2}, \\ 0 \leq x^2 + y^2 < 1 \end{cases} \tag{5.61}$$

Therefore, to find the geodesic lines S on the disc $D^2 : 0 \leq x^2 + y^2 < 1$ for the metric form (5.50), we should now consider the following system of equations:

$$\begin{cases} X = \dfrac{2x}{1-x^2-y^2}, \\ Y = \dfrac{2y}{1-x^2-y^2}, \\ Z = \dfrac{1+x^2+y^2}{1-x^2-y^2}, \\ aX + bY + cZ = 0. \end{cases} \tag{5.62}$$

Then, the equation of the **geodesic lines** $S(x, y) \in D^2 : 0 \leq x^2 + y^2 < 1$ for the metric form (5.50), after substitution $X = \dfrac{2x}{1-x^2-y^2}$, $Y = \dfrac{2y}{1-x^2-y^2}$, $Z = \dfrac{1+x^2+y^2}{1-x^2-y^2}$ into the equation $aX + bY + cZ = 0$ and reducing the common factor $\dfrac{1}{1-x^2-y^2}$, takes the following form:

$$S : 2ax + 2by = -c\left(1 + x^2 + y^2\right). \tag{5.63}$$

Let us consider two cases: 1) $c=0$; 2) $c \neq 0$.

For the case 1) $c=0$, the equation (5.63) for the **geodesic lines** S on the disc $D^2 : 0 \le x^2 + y^2 < 1$ takes the following form: $2ax + 2by = 0 \Rightarrow ax + by = 0$, that is, the geodesic lines S are the line segments, belonging to the disc $D^2 : 0 \le x^2 + y^2 < 1$.

For the case 2) $c \ne 0$, the equation (5.63) takes the following form:

$$\begin{cases} -\dfrac{a}{c}2x - \dfrac{b}{c}2y = 1 + x^2 + y^2 \Rightarrow \\ \Rightarrow 1 + x^2 + y^2 = 2x_0 x + 2y_0 y \Rightarrow \\ \Rightarrow x^2 - 2x_0 x + y^2 - 2y_0 y = -1 \Rightarrow \\ \Rightarrow x^2 - 2x_0 x + x_0^2 + y^2 - 2y_0 y + y_0^2 = -1 + x_0^2 + y_0^2 \Rightarrow \\ \Rightarrow (x - x_0)^2 + (y - y_0)^2 = \rho^2 \end{cases} \qquad (5.64)$$

where $x_0 = -\dfrac{a}{c}, y_0 = -\dfrac{b}{c}$, $\rho = \sqrt{x_0^2 + y_0^2 - 1}$.

Consequently, the **geodesic lines** on S for the case $c \ne 0$ are arcs of the circles

$$(x - x_0)^2 + (y - y_0)^2 = \rho^2, \qquad (5.65)$$

belonging to the disc $D^2 : 0 \le x^2 + y^2 < 1$. Note that the center (x_0, y_0) of the circles (5.65) is outside of the disc $D^2 : 0 \le x^2 + y^2 < 1$. Direct verification establishes that the circles (5.65) intersect with the **absolute** $x^2 + y^2 = 1$ under the right angle.

5.6.2. The relationship between Poincaré's metric model and Lobachevski's metric (λ, μ)-models

Assume that $\alpha = \ln(\Phi_\lambda), \beta = \ln(\Phi_\mu), (\lambda, \mu) > 0$, then, we get:

$$sh(\alpha u') = \frac{\Phi_\lambda^{u'} - \Phi_\lambda^{-u'}}{2},$$

$$ch(\alpha u') = \frac{\Phi_\lambda^{u'} + \Phi_\lambda^{-u'}}{2},$$

$$th(\alpha u') = \frac{\Phi_\lambda^{u'} - \Phi_\lambda^{-u'}}{\Phi_\lambda^{u'} + \Phi_\lambda^{-u'}},$$

$$cth(\alpha u') = \frac{\Phi_\lambda^{u'} + \Phi_\lambda^{-u'}}{\Phi_\lambda^{u'} - \Phi_\lambda^{-u'}},$$

$$\sin(\beta u') = \sin\left(\ln(\Phi_\mu)u'\right),$$

$$\cos(\beta u') = \cos\left(\ln\left(\Phi_\mu\right)u'\right).$$

Hence, the formulas (5.52) - (5.55), obtained after the recalculation in the terms of the "metallic proportions" and "hyperbolic Fibonacci goniometry," between Poincaré's metric model and Lobachevski's metric (λ,μ)-models, take the following forms:

$$\begin{cases} \dfrac{\Phi_\lambda^{u'} - \Phi_\lambda^{-u'}}{2} = \dfrac{2\sqrt{x^2 + y^2}}{1 - \left(x^2 + y^2\right)} \\[4mm] \dfrac{\Phi_\lambda^{u'} + \Phi_\lambda^{-u'}}{2} = \dfrac{1 + \left(x^2 + y^2\right)}{1 - \left(x^2 + y^2\right)} \end{cases} \tag{5.66}$$

$$\begin{cases} \sin\left(\ln\left(\Phi_\mu\right)u'\right) = \dfrac{y}{\sqrt{x^2 + y^2}} \\[4mm] \cos\left(\ln\left(\Phi_\mu\right)u'\right) = \dfrac{x}{\sqrt{x^2 + y^2}} \end{cases} \tag{5.67}$$

$$y = y(u',v') = \sqrt{\dfrac{\Phi_\lambda^{u'} - \Phi_\lambda^{-u'} - 2}{\Phi_\lambda^{u'} + \Phi_\lambda^{-u'} + 2}}\,\sin\left(\ln\left(\Phi_\mu\right)v'\right), \tag{5.68}$$

$$(ds)^2 = \frac{4\left[(dx)^2 + (dy)^2\right]}{\left(1 - x^2 - y^2\right)^2} = \ln^2\left(\Phi_\lambda\right)(du')^2 + \ln^2\left(\Phi_\mu\right)sh^2\left[\ln\left(\Phi_\lambda\right)u'\right](dv')^2 = (ds')^2. \tag{5.69}$$

5.7. The basic results of the study

5.7.1 The original solution (pseudo-spherical solution) of Hilbert's Fourth Problem

Thus, Lobachevski's metric (λ,μ)-forms (5.37) give a countless set of geometries, based on the "metallic proportions" and "hyperbolic Fibonacci goniometry." On the one hand, they possess the properties, similar to Lobachevski's classical metric form (5.8) (preservation of the same Gaussian curvature $K < 0$, isometric equivalence, equiareality), on the other hand, when compared with Lobachevski's classical metric form (5.8), they have their specific properties (isometry, non-conformity).

For the case $\lambda, \mu \neq 1$, Lobachevski's metric (λ,μ)-forms (5.37) are on the finite distance $\bar{\rho}_{12} > 0$ from Lobachevski's metric (λ,μ)-forms (5.37), moreover, for the case $\lambda,\mu \to 1$, the distance $\bar{\rho}_{12} \to 0$. **This means that for the case $\lambda,\mu \to 1$ Lobachevski's metric (λ,μ)-forms (5.37) are arbitrarily close to Lobachevski's metric form (5.8) and coincide with (5.8) for the case $\lambda = \mu = 1$.**

It is clear that these new Lobachevski's geometries *"with equal right, stand next to Euclidean geometry"* (**David Hilbert**). It follows from this consideration that **the formula (5.37) can be considered as the partial original solution to Hilbert's Fourth Problem.** That is, there is an infinite number of Lobachevski's geometries, described by the formula (5.37), which are close to Euclidean geometry.

Each of these new Lobachevski's geometries, based on the "metallic proportions" and "hyperbolic Fibonacci goniometry," can actually exist in the real physical world, similarly to "Bodnar's geometry" [20], which is a unique hyperbolic geometry of botanical phenomenon of phyllotaxis, widely common in wildlife. As it is proved in [20], the hyperbolic geometry of phyllotaxis is based on the classical "golden ratio" and Fibonacci numbers, which appear at the surface of phyllotaxis objects.

5.7.2. A new challenge for theoretical natural sciences

Thus, the main result of the research, described in the present Chapter, is a proof of the existence of an infinite number of hyperbolic functions – Fibonacci hyperbolic λ-functions, based on the "metallic proportions." For the given $\lambda=1,2,3,..$ each class of the Fibonacci hyperbolic λ-functions, "generates" new hyperbolic geometry, which leads to the appearance in the "physical world" specific hyperbolic geometries with properties, which depend on the "metallic means." The new geometric theory of phyllotaxis, created by Oleg Bodnar [20], is a striking example of this. Bodnar proved that "the world of phyllotaxis" is a specific "hyperbolic world," in which a "hyperbolicity" manifests itself in the "Fibonacci spirals" on the surface of "phyllotaxis objects."

Recall that "Bodnar's geometry" [20] is based on the Fibonacci hyperbolic functions, also known as the **"golden" hyperbolic functions**:

$$\begin{cases} sFs(x) & = & \dfrac{\Phi^{x} - \Phi^{-x}}{\sqrt{5}} \\[4mm] cFs(x) & = & \dfrac{\Phi^{x} + \Phi^{-x}}{\sqrt{5}} \end{cases} \qquad (5.70)$$

which are connected with Fibonacci numbers.

The famous irrational number $\Phi = \dfrac{1+\sqrt{5}}{2} \approx 1.618$ is the base of the functions (5.70). The distance of "Bodnar' geometry" to Lobachevski's geometry is equal $\overline{\rho}_{12} \approx 0.7336$.

However, the "golden" hyperbolic functions (5.70), which underlie the "hyperbolic phyllotaxis world," are a special case of the hyperbolic Fibonacci λ - functions (λ=1). In this regard, there is every reason to suppose that other types of hyperbolic functions - **Fibonacci hyperbolic λ-functions** - can be the basis for modeling of the new "hyperbolic worlds," which can really exist in Nature. Modern science could not find these special "hyperbolic worlds," because the Fibonacci hyperbolic λ-functions were unknown until to the works [13,14]. For the first time, these hyperbolic functions are described in Stakhov's articles [13,14] and in Stakhov's and Aranson's articles [29-31]. Basing on the success of "Bodnar's geometry" [20], one can put forward in front to theoretical physics, chemistry, crystallography, botany, biology, and other branches of theoretical natural sciences **the challenge to search new ("harmonic") hyperbolic worlds of Nature, based on other classes of the Fibonacci hyperbolic λ-functions** (4.75), (4.76).

In this case, perhaps, the first candidate for the new "hyperbolic world" of Nature may be, for example, **"silver" hyperbolic functions**:

$$\begin{cases} sF_2(x) = \dfrac{\Phi_2^x - \Phi_2^{-x}}{\sqrt{8}} = \dfrac{1}{2\sqrt{2}}\left[\left(1+\sqrt{2}\right)^x - \left(1+\sqrt{2}\right)^{-x}\right] \\ cF_2(x) = \dfrac{\Phi_2^x + \Phi_2^{-x}}{\sqrt{8}} = \dfrac{1}{2\sqrt{2}}\left[\left(1+\sqrt{2}\right)^x + \left(1+\sqrt{2}\right)^{-x}\right] \end{cases}, \tag{5.71}$$

which are connected with Pell numbers and are based on the "silver mean" $\Phi_2 = 1+\sqrt{2} \approx 2.41$, connected with the fundamental mathematical constant $\sqrt{2}$.

In this regard, we should draw a special attention to the fact that the new hyperbolic geometry, based on the "silver" hyperbolic functions (5.71), is the closest to Lobachevski's geometry, based of the classical hyperbolic functions with the base $e \approx 2.71$. Its distance to Lobachevski's geometry is equal $\overline{\rho}_{12} \approx 0.1677$ what is the smallest among all the distances for Lobachevski's metric (λ,μ)-forms. It allows to assume that the "silver" hyperbolic functions (5.71) and the generated by them "silver" hyperbolic geometry can be soon be found in Nature after "Bodnar's geometry," based on the "golden" hyperbolic functions (5.70).

Chapter 6

SPHERICAL FIBONACCI FUNCTIONS AND SPHERICAL SOLUTION OF HILBERT'S FOURTH PROBLEM

6.1. Basic metric forms of the spherical geometry

6.1.1. Preliminaries

In this chapter from the point of view of mathematics of harmony, we will study the geometric and metric properties of two-dimensional sphere not only as a surface in three-dimensional Euclidean space, but as a two-dimensional manifold associated with its various parameterizations.

Comparison of these properties with Lobachevski's geometry, induced by a similar study of the pseudo-sphere of Minkowski's three-dimensional space (see Chapter 5), as a result, allows the authors to talk about new solution of Hilbert's Fourth Problem.

General concepts about the history of spherical geometry and a modern look on this geometry as on the non-Euclidean geometry with a positive Gaussian curvature are presented, for example, in [4], [5].

The spherical geometry has not only theoretical importance but also many practical applications to important branches of natural science, in particular, cartography, navigation, astronomy, and so on.

6.1.2. Standard spherical metric forms

A sphere S^2 of the radius $R>0$ with the center $(0,0,0)$ in the Euclidean space $R^3=(X, Y, Z)$ is given by the equation:

$$S^2: \quad X^2+Y^2+Z^2= R^2. \tag{6.1}$$

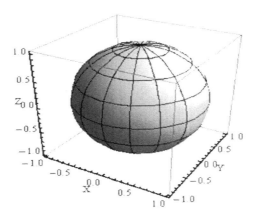

Figure 6.1. Sphere S^2: $X^2+Y^2+Z^2=R^2$

In spherical coordinates the sphere S^2 admits the following parameterization:

$$\begin{cases} X = R\sin(u)\cos(v), \\ Y = R\sin(u)\sin(v), \\ Z = R\cos(u), \end{cases} \tag{6.2}$$

where the parameters u,v satisfy to the condition:

$$-\infty < u,v < +\infty. \tag{6.3}$$

We use the following names for the sphere S^2:

north pole N: $(X=0,Y=0,Z=R) \Leftrightarrow (u = 2\pi k, -\infty < v < +\infty)$,

south pole S: $(X=0,Y=0,Z=-R) \Leftrightarrow (u = \pi(2k+1), -\infty < v < +\infty)$,

equator Q: $(X^2+Y^2=R^2, Z=0) \Leftrightarrow (u = \dfrac{\pi}{2}(2k+1), -\infty < v < +\infty)$,

where u is *zenith angle, v is azimuth angle,* $k = 0,\pm1,\pm2,....$.

In spherical coordinates (u, v) under the additional restriction $u \neq \pi k$, that is, for the case $\pi k < u < \pi(k+1)$, $-\infty < v < +\infty$ $(k = 0,\pm1,\pm2,...)$, the metric form of the sphere S^2 has the following form:

$$(ds)^2 = R^2\left[(du)^2 + \sin^2(u)(dv)^2\right], \tag{6.4}$$

$$E = R^2 > 0,\ F = 0,\ G = R^2 \sin^2(u) > 0,\ EG - F^2 = R^4 \sin^2(u) > 0,$$

where ds is an element of arc length on the plane (u,v).

Let us name the metric form (6.4) for each fixed value of the radius $R>0$ of the sphere S^2 of kind (6.1) **standard spherical metric form**.

In the literature for the sphere S^2 instead of the metric form (6.4) we use often the form of the kind $(ds)^2 = (dx)^2 + \cos^2(\frac{x}{R})(dv)^2$ of the same Gaussian curvature $K = \frac{1}{R^2} > 0$, as for the form (6.4). This form is obtained from the form

(6.4) after the transformation of variables $u = \dfrac{\pi}{2} - \dfrac{x}{R}, v = \dfrac{y}{R}$. Hence we have:

$$(dl)^2 = (ds)^2 = R^2 \left[(du)^2 + \sin^2(u)(dv)^2 \right] = (dx)^2 + \cos^2\left(\frac{x}{R}\right)(dy)^2,$$

where $(dl)^2 = (dX)^2 + (dY)^2 + (dZ)^2$ is Euclidean metrics in the space (X,Y,Z). The

transformation $u = \dfrac{\pi}{2} - \dfrac{x}{R}, v = \dfrac{y}{R}$ induces a special case of **sliding (overlay)** of the sphere on itself.

The famous Italian mathematician **Eugenio Beltrami** (1835-1900) established the following general result for surfaces of the constant Gaussian curvature: every surface of the constant curvature can move on itself (at least by **bending**), with three degrees of freedom.

For the sphere and the plane, such transformations are a set of three transformations: **rotations, reflections** and **compressions (expansions)**. In the future, these results have been developed and refined.

Stereographic projection from the *north pole* $N(0,0,R)$ of the sphere S^2 on the plane $\Pi:Z=0$, in the coordinates $(x,y) \in \Pi$ has the following form:

$$X = \frac{2R^2 x}{x^2 + y^2 + R^2},\ Y = \frac{2R^2 y}{x^2 + y^2 + R^2},\ Z = R\frac{x^2 + y^2 - R^2}{x^2 + y^2 + R^2} \qquad (6.5)$$

<u>Proof.</u> Let us consider the stereographic projection from the *north pole* $N(0,0,R) \in S^2$. For this purpose, in the space R^3 through two points: $N(X_1=0, Y_1=0, Z_1=R) \in S^2$, $P(X_2=x, Y_2=y, Z_2=0) \in \Pi$, we conduct the straight L: $\dfrac{X-0}{x-0} = \dfrac{Y-0}{y-0} = \dfrac{Z-R}{0-R}$, and then we find the point (X,Y,Z) on the intersection of the straight L with the sphere S^2: $X^2 + Y^2 + Z^2 = R^2$.

Let us rewrite the equation of the straight L in parametric form:

$$L: X=xt,\ Y=yt,\ Z=R(1-t),\ -\infty < t < +\infty. \qquad (6.6)$$

Substituting (6.6) into the equation of the sphere (6.1) S^2: $X^2 + Y^2 + Z^2 = R^2$, we get two roots:

$$t_1 = 0, t_2 = \frac{2R^2}{x^2 + y^2 + R^2} \tag{6.7}$$

For the case of the root $t_1 = 0$ at all $-\infty < x, y < +\infty$ we get: $X_1 = 0$, $Y_1 = 0$, $Z_1 = R$, that the value $t_1 = 0$ corresponds to the *north pole* $N(0, 0, R) \in S^2$.

For the root $t_2 = \dfrac{2R^2}{x^2 + y^2 + R^2}$ at all $-\infty < x, y < +\infty$ we get:

$$X = \frac{2R^2 x}{x^2 + y^2 + R^2}, Y = \frac{2R^2 y}{x^2 + y^2 + R^2}, Z = R\frac{x^2 + y^2 - R^2}{x^2 + y^2 + R^2}.$$

Let us introduce the following designation: $r = \sqrt{x^2 + y^2}$. Then, we have:

$$X^2 + Y^2 = 4R^4 \frac{r^2}{\left(r^2 + R^2\right)^2}, \quad Z = R\frac{r^2 - R^2}{r^2 + R^2}. \tag{6.8}$$

Hence, we obtain:

1). For $r = \sqrt{x^2 + y^2} \to +\infty$, $X^2 + Y^2 \to 0, Z \to R$ (*north pole*);

2). For $r = R$, $x^2 + y^2 = R^2$, $X^2 + Y^2 = R^2, Z = 0$ (*equator*);

3). При $r = 0$, $x = 0, y = 0$, $X = 0, Y = 0, Z = -R$ (*south pole*).

For the coordinates $(x, y,)$, the metric form of the sphere (1) has the following form:

$$\left(ds\right)^2 = \left(dl\right)^2 = \frac{4R^2}{\left(x^2 + y^2 + R^2\right)^2}\left[\left(dx\right)^2 + \left(dy\right)^2\right], \quad E = G = \frac{4R^2}{\left(x^2 + y^2 + R^2\right)^2} > 0. \tag{6.9}$$

We call the metric form (6.9) the ***normal spherical metric form***.

6.1.3. Absolutes for the standard and normal spherical metric forms

The straights $u = \pi k, -\infty < v < +\infty$ are the *absolutes* for the standard spherical metric form (6.4), that is, the *north* and *south* poles $X = 0, Y = 0, Z = \pm R$ for the sphere S^2 because for the case $u = \pi k, -\infty < v < +\infty$ we get the degeneration of the coefficient G:

$$E = R^2 > 0, F = 0, G = R^2 \sin^2\left(\pi k\right) = 0.$$

2. For the normal metrical form (1.9), the *infinity circle* $x^2 + y^2 = +\infty$ is the *absolute*, that is, the *north pole* $X = 0, Y = 0, Z = R$ for the sphere S^2, because at $x^2 + y^2 = +\infty$ we get the degeneration of the coefficients G, E:

$$E = G = \frac{4R^2}{\left(x^2 + y^2 + R^2\right)^2} \to 0, F = 0.$$

6.1.4. The one-to-one mapping between the coordinates (u,v) and (x,y) for the sphere S^2 without the north and south poles

In this section, unless otherwise stated, we will consider the coordinates (u, v) and (x, y) of the sphere $S^2 : X^2 + Y^2 + Z^2 = R^2$ without the *north* pole N ($X=0, Y=0$, $Z=R$) and the *south* pole $S(X=0, Y=0, Z=R$). Then, in the coordinates (u, v) the allowed values are: $u \neq \pi k, -\infty < v < +\infty$, and in the coordinates (x, y) the allowed values are: $0 < r = \sqrt{x^2 + y^2} < +\infty$.

By virtue of (6.2) and (6.5), we have the following equations:

$$X = R\sin(u)\cos(v) = \frac{2R^2 x}{r^2 + R^2}, Y = R\sin(u)\sin(v) = \frac{2R^2 y}{r^2 + R^2}, Z = R\cos(u) = R\frac{r^2 - R^2}{r^2 + R^2}.$$

Hence, we obtain the following one-to-one mapping between the coordinates (u, v) and (x, y).

1. The transformation of the kind: $(x,y) \mapsto (u,v)$.

$$\sin(u) = \frac{2Rr}{r^2 + R^2}, \ \cos(u) = R\frac{r^2 - R^2}{r^2 + R^2}, \ \sin(v) = \frac{y}{r}, \ \cos(v) = \frac{x}{r}, \ r = \sqrt{x^2 + y^2}. \quad (6.10)$$

2. The transformation of the kind: $(u,v) \mapsto (x, y)$.

$$x = R\sqrt{\frac{1 + \cos(u)}{1 - \cos(u)}}\cos(v), \ y = R\sqrt{\frac{1 + \cos(u)}{1 - \cos(u)}}\sin(v). \quad (6.11)$$

6.1.5. Geodesic lines for the standard and normal metric spherical forms

Geodesic lines L on the sphere $S^2 : X^2 + Y^2 + Z^2 = R^2$ in the Euclidean space R^3 $=(X, Y, Z)$ are the circles of big discs, which are obtained as a result of the intersection of the sphere S^2 with the planes Π: $aX + bY + cZ = 0$ $(a^2 + b^2 + c^2 > 0)$, passing through the coordinates origin $O(0,0,0)$. Hence, to find the geodesic lines $L = S^2 \cap \Pi$, we get the system of the equation:

$$\begin{cases} aX + bY + cZ = 0 \\ X^2 + Y^2 + Z^2 = R^2 \end{cases} \qquad (6.12)$$

6.1.5.1. Geodesic lines for the standard spherical metric form (6.4)

In the Section 6.1.5.1 we consider the sphere $S^2 : X^2 + Y^2 + Z^2 = R^2$ without the *north* pole N ($X=0, Y=0, Z=R$) and the *south* pole S ($X=0, Y=0, Z=-R$), what means for the coordinates (u,v) that $u \neq \pi k$, where $k = 0, \pm1, \pm2, \dots$. In this case, the standard spherical metric form has the form (6.4), and the parameterization of the sphere S^2 has the form (6.2).

There are two cases: 1) $c=0$; 2) $c \neq 0$.

For the case 1) $c=0$, we get the following relations:

$$aX + bY = 0 \Rightarrow aR\sin(u)\cos(v) + bR\sin(u)\sin(v) = 0. \qquad (6.13)$$

Because $u \neq \pi k, -\infty < v < +\infty$, then we get from (6.13) the equation of the geodesic lines L in the form: $a\cos(v) + b\sin(v) = 0$, $a^2 + b^2 > 0$. It follows from here:

$$L: \ \pi k < u < \pi(k+1), \ v = const, \ k = 0, \pm1, \pm2, \dots. \qquad (6.14)$$

For the case 2) $c \neq 0$, we get the following relations:

$$Z = AX + BY, \ A = -\frac{a}{c}, \ B = -\frac{b}{c} \ \Rightarrow R\cos(u) = AR\sin(u)\cos(v) + BR\sin(u)\sin(v) \Rightarrow$$

$$\Rightarrow \cos(u) = \sin(u)[A\cos(v) + B\sin(v)].$$

Because $u \neq \pi k$, then the equation of the geodesic lines has the following form:

$$L: \ ctg(u) = A\cos(v) + B\sin(v), \ \pi k < u < \pi(k+1), \ -\infty < v < +\infty, \ k = 0, \pm1, \pm2, \dots. \qquad (6.15)$$

6.1.5.2 Geodesic lines for the normal spherical metric form (6.9)

In the Section 6.1.5.2 we consider the sphere $S^2 : X^2 + Y^2 + Z^2 = R^2$ without the *north pole* N ($X=0, Y=0, Z=R$), but at the presence of the *south pole* S ($X=0, Y=0, Z=-R$), what means in the coordinates (x,y), that $0 \leq r = \sqrt{x^2 + y^2} < +\infty$.

For this case, the normal spherical metric form has the form (6.9), and the parameterization of the sphere S^2 has the form (6.5). The plane has the form: Π: $aX+bY+cZ=0$, where $a^2+b^2+c^2>0$. Substituting (6.5) into the equation of the plane Π, we get that in the coordinates (x,y) the geodesic lines have the following form:

$$L:\ a(2R)x+b(2R)y=-c\left(x^2+y^2-R^2\right).\tag{6.16}$$

There are two cases: 1) $c=0$; 2) $c\neq 0$.

For the case 1) $c=0$, we get the following relations:

$$L:\ ax+by=0,\ a^2+b^2>0.\tag{6.17}$$

For the case 2) $c\neq 0$, we use the following designation $x_0=-\dfrac{a}{c}$, $y_0=-\dfrac{b}{c}$.

Then, in the coordinates $(x,\ y)$, we get that geodesic lines are the circles, having the form:

$$L:\ L:(x-Rx_0)^2+(y-Ry_0)^2=R^2\left(x_0^2+y_0^2+1\right).\tag{6.18}$$

The point $(Rx_0,\ Ry_0)$ is the center of the circle (6.18), the radius is equal to $\rho_0=R\sqrt{x_0^2+y_0^2+1}$.

It is established directly that the coordinates origin $O(x=0,\ y=0)$ for any $-\infty<x_0,y_0<+\infty$ belongs to the disc $(x-Rx_0)^2+(y-Ry_0)^2<R^2\left(x_0^2+y_0^2+1\right)$ and two any circles L and L' of the kind (6.18) are intersected in two different points.

6.1.6. The motions of the sphere S^2 in the coordinates (x, y)

In the Section 6.1.6, we consider the sphere S^2:$X^2+Y^2+Z^2=R^2$ without the *north pole*
N ($X=0,Y=0,\ Z=R$), but at the presence of the *south pole* S ($X=0,Y=0,\ Z=-R$), what means in the coordinates $(x,y$), that $0\leq r=\sqrt{x^2+y^2}<+\infty$.

In the complex form, the metric form (6.9) is written as follows:

$$(dl)^2=\frac{4R^2}{\left(|z|^2+R^2\right)^2}\,dz\,d\bar z,\ z=x+iy,\bar z=x-iy,|z|=\sqrt{x^2+y^2}.\tag{6.19}$$

The *motions* of the sphere S^2 are a combination of two kinds of transformations: *reflections* and *rotations*. The *reflections* change the orientation of the complex z-plane and have the form: $z\mapsto\bar z$. The *rotations* preserve the orientation of the complex *z*-plane and are the *motions* of the metric form (6.19), preserving the conformal form of this metric form. This means that the motions of the metric form (6.19) preserve not only the elements of the arc lengths dl, but the angles between the arcs.

The motions of the metric form (6.19) has the following form [5]:

$$z = z(w) = \frac{Aw + B}{Cw + D}, \quad AD\text{-}BC = 1, \tag{6.20}$$

where

$$(dl)^2 = \frac{4R^2}{\left(|z|^2 + R^2\right)^2} dz d\bar{z} = \frac{4R^2}{\left(|w|^2 + R^2\right)^2} dw d\bar{w} \tag{6.21}$$

We make the change of the variables:

$$z = Rz', \, w = Rw' \Rightarrow z' = \frac{z}{R}, \, w' = \frac{w}{R} \tag{6.22}$$

Then, we get from (6.21) and (6.22):

$$(dl)^2 = \frac{4R^2}{\left(|z'|^2 + 1\right)^2} dz d\bar{z} = \frac{4R^2}{\left(|w'|^2 + 1\right)^2} dw' d\bar{w'} \tag{6.23}$$

It is known that the *motion* of the metric form (6.23) are realized by means of the linear-fractional transformation:

$$z' = z'(w') = \frac{aw' + b}{cw' + d}, \tag{6.24}$$

where the coefficients *a,b,c,d* satisfy to the following equalities:

$$ad - bc = 1, \, |a|^2 + |c|^2 = 1, \, |b|^2 + |d|^2 = 1, \, a\bar{b} + c\bar{d} = 0 \tag{6.25}$$

Then, we get:

$$z' = \frac{z}{R} = \frac{aw' + b}{cw' + d} = \frac{a\frac{w}{R} + b}{c\frac{w}{R} + d} \Rightarrow \frac{z}{R} = \frac{aw + bR}{cw + dR} \Rightarrow z = \frac{aRw + bR^2}{cw + dR} \equiv \frac{Aw + B}{Cw + D}. \tag{6.26}$$

Since *AD-BC*=1, then we can introduce the normalization factor *t*> 0 into the relation (6.26), then, the formula (6.26) can be rewritten as follows:

$$z = \frac{(taR)w + tbR^2}{(tc)w + tdR} \equiv \frac{Aw + B}{Cw + D} \tag{6.27}$$

From here, it follows the relations:

196

$$A = Rat, \quad B = bR^2t, \quad C = ct, \quad D = dRt \qquad (6.28)$$

Substituting (6.28) into the relation $AD-BC=1$, we get:

$$t^2 = \frac{1}{R^2(ad-bc)}. \qquad (6.29)$$

According to the conditions (6.25), $ad-bc=1$, therefore for the case $t>0$ we get from (6.29): $t = \frac{1}{R}$.

From here, the relations (6.28) can be rewritten as follows:

$$A=a, B=bR, \ C=\frac{c}{R}, D=d \Rightarrow a=A, b=\frac{B}{R}, c=CR, d=D. \qquad (6.30)$$

Substituting (6.30) into (6.25), we get:

$$AD-BC=1, \quad |A|^2 + R^2|C|^2 = 1, \quad |B|^2 + R^2|D|^2 = R^2, \quad A\bar{B}+R^2C\bar{D}=0. \quad (6.31)$$

6.1.7. The distance on the sphere S^2 in the coordinates (u,v) with regard for the north and south poles

The **distance** L ($0 \le L \le \pi R$) on the sphere S^2 can be given as follows:

$$L = R \bullet \arccos(m), \ m = \cos(u_1)\cos(u_2) + \sin(u_1)\sin(u_2)\cos(v_1 - v_2), \qquad (6.32)$$

where $-\infty < u_1, v_1, u_2, v_2 < +\infty, |m| \le 1$.

The maximum possible value of $L=\pi R$ is equal to the distance between the *north pole* $N:(u_1 = 2\pi k, -\infty < u_1 < +\infty)$ and the *south pole* $S:(u_2 = \pi(2k+1), -\infty < v_2 < +\infty)$, and also between any of the other diametrically opposed points

$A_1:(-\infty < u_1 < +\infty, -\infty < v_1 < +\infty)$ and $A_2:(u_2 = u_1(2k+1)\pi, -\infty < v_2 = v_1 + 2\pi k < +\infty)$ of the sphere S^2.

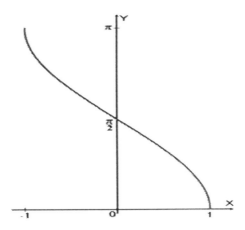

Figure 6.2. The graph of the function $Y = \arccos(x)$, *where* $x=m$, $Y = \dfrac{L}{R}$

6.2. The perturbed spherical metric forms

6.2.1. The parameterized spherical metric forms

Along with the standard spherical metric form (6.4) of the sphere S^2 of the form (6.1) with the radius $R>0$, for which $k\pi < u < (k+1)\pi, -\infty < v < +\infty$, we will consider the spherical metric forms of the sphere S^2 of the form (6.1) with the radius $R>0$, which depend on the real parameter $\alpha > 0$. These forms are looked as follows:

$$(ds')^2 = R^2\left[(\alpha^2 du')^2 + \sin^2(\alpha u')(dv')^2\right],\tag{6.33}$$

where $E' = R^2\alpha^2 > 0, F' = 0, G' = R^2\sin^2(\alpha u') > 0$ are the metric coefficients of the form (6.33).

We call the forms (6.33) *parameterized spherical metric forms*, and each real parameter $\alpha > 0$ a *parameterization parameter*.

For each fixed value of $\alpha > 0$, the straights $u' = \dfrac{k}{\alpha}\pi$, $-\infty < v' < +\infty$, $k \in Z$ (Z is the set of integers) are the *absolutes* for the parameterized spherical metric form (6.33), because on these straights the metric coefficient $G' = R^2\sin^2(\pi k) = 0$, that is, there is a *degeneration* of at least one of the metric coefficients of the form (6.33). Therefore, the *non-degenerate* parameterized spherical metric forms

(6.33) for the variables $(u'v')$ with the parameters of parameterization of $\alpha > 0$ exist only for the following restrictions:

$$\frac{k}{\alpha}\pi < u' < \frac{k+1}{\alpha}\pi, -\infty < v' < +\infty, k \in Z. \tag{6.34}$$

6.2.2. Gaussian curvature for the spherical metric parameterized forms

In this section we consider the parameterized spherical metric form (6.33). Let us show that for the condition $u' \neq \dfrac{k}{\alpha}\pi$, $-\infty < v' < +\infty$ $(k \in Z)$, $\alpha > 0$ the *Gaussian curvature* $K' = \dfrac{1}{R^2} > 0$.

It is known [5], if the metric form has the form

$$(ds')^2 = E'(u',v')(du')^2 + G'(u',v')(dv')^2, E' = E'(u',v') > 0, G' = G'(u',v') > 0, \tag{6.35}$$

then the Gaussian curvature is calculated by the formula:

$$K = -\frac{1}{A'B'}\left(\frac{\partial}{\partial u'}\left(\frac{\frac{\partial B'}{\partial u'}}{A'}\right) + \frac{\partial}{\partial v'}\left(\frac{\frac{\partial A'}{\partial v'}}{B'}\right)\right), A' = \sqrt{E'}, B' = \sqrt{G'}. \tag{6.36}$$

For our situation we have:

$$E' = R^2\alpha^2 > 0, G' = R^2\sin^2(\alpha u') > 0, A' = \sqrt{E'} = R\alpha > 0,$$

$$B' = R\sqrt{\sin^2(\alpha u')} = R|\sin(\alpha u')| > 0, \frac{\partial^2 B'}{(\partial u')^2} = -R\alpha^2\sqrt{\sin^2(\alpha u')} = -R\alpha^2|\sin(\alpha u')| > 0.$$

Hence, the Gaussian curvature is given by:

$$K = K(u';\alpha) = -\frac{1}{(A')^2 B'}\frac{\partial^2 B'}{(\partial u')^2} = [-\frac{1}{R^3\alpha^2|\sin(\alpha u')|}]\bullet[-R\alpha^2|\sin(\alpha u')|] = \frac{1}{R^2} > 0.$$

Note, if $u' \to \dfrac{k}{\alpha}\pi$, we get: $K(u';\alpha) \to \dfrac{1}{R^2} > 0$.

When the value of the weight coefficient $\alpha = 1$, the parameterized metric form (6.33) coincides with the standard spherical form (6.4) for the case of the identity mapping $u \equiv u', v \equiv v'$.

But then the Gaussian curvature of the standard spherical metric form (6.4) is also equal to $K = \dfrac{1}{R^2} > 0$. Similarly, we can find that in the coordinates (x, y) for

199

the normal spherical metric form (6.9), the Gaussian curvature is also equal to

$$K = \frac{1}{R^2} > 0.$$

6.2.3. Normalized distance between the parameterized spherical metric forms and standard spherical metric form

For each fixed value $R>0$ we can introduce the **distance** ρ_{12} between the parameterized spherical metric forms (6.33), where $\frac{k}{\alpha}\pi < u' < \frac{k+1}{\alpha}\pi, -\infty < v' < +\infty$, $k \in Z$, $\alpha > 0$, and standard spherical metric form (6.4) as follows:

$$\rho_{12} = R|(\alpha - 1|) . \tag{6.37}$$

In the future, unless otherwise stated, we will consider only the **normalized distance** $\overline{\rho}_{12}$ between the parameterized spherical metric forms (6.33) and the standard metric spherical form (6.4):

$$\overline{\rho}_{12} = \frac{\rho_{12}}{R} = |\alpha - 1| . \tag{6.38}$$

Note that $\overline{\rho}_{12}$ does not depend on the radius R of the sphere (6.1).

If $\overline{\rho}_{12} > 0 \Leftrightarrow \alpha > 0, \alpha \neq 1$, then the parameterized spherical metric form (6.33) do not coincide with the standard spherical metric form (6.4).

If $\overline{\rho}_{12} = 0 \Leftrightarrow \alpha = 1$, then the parameterized spherical metric form (6.33) coincide with the standard spherical metric form (6.4), thus we should believe $u = u', v = v'$.

6.2.4. Comparison of the metric properties of the parameterized spherical metric forms to the standard spherical metric form for the case $\overline{\rho}12 > 0$

6.2.4.1. The common metric properties

We list the metric properties, which at the change of the variables

$$f: u = \alpha u', v = v' \ (\alpha > 0) \tag{6.39}$$

between the standard spherical metric form (6.4) ($\overline{\rho}_{12} = 0$) and the parameterized spherical metric forms (6.33) ($\overline{\rho}_{12} > 0$) are the same (saved).

200

I. Saving the Gaussian curvature

It was shown above that the preservation of the Gaussian curvature $K = \dfrac{1}{R^2} > 0$, where R is the radius of the sphere (6.1), is a common property both for the standard spherical metric form (6.4) ($\overline{\rho}_{12} = 0$), and for the parameterized spherical metric forms (6.33) ($\overline{\rho}_{12} > 0$).

II. Isometric equivalence

Suppose that in the areas $D = (u, v)$ and $D' = (u', v')$ the two metric forms are given, respectively, as follows:

$$(ds)^2 = E(u,v)(du)^2 + 2F(u,v)(dudv) + G(u,v)(dv)^2, \ E > 0, G > 0, EG - F^2 > 0, \tag{6.40}$$

$$(ds')^2 = E'(u',v')(du')^2 + 2F'(u',v')(du'\,dv') + G'(u',v')(dv')^2, E' > 0, G' > 0, E'G' - (F')^2 > 0. \tag{6.41}$$

Suppose that the diffeomorphism of the kind $f : D \mapsto D'$ is given as follows:

$$f: \ u = u(u', v'). \ v = v(u', v'), \frac{\partial(u, v)}{\partial(u', v')} \neq 0 \tag{6.42}$$

According to the **Definition** 5.1, the metric forms (6.4) and (6.33) are **isometrically equivalent** under the action of the diffeomorphism (6.42) $\overline{\rho}_{12} > 0$; this means the fulfilment of the identities (5.14), whence it follows the coincidence of the arc lengths elements:

$$ds = ds'. \tag{6.43}$$

We show that for our situation, the standard spherical form (6.4) and the parameterized metric form (6.33) for the case $\overline{\rho}_{12} > 0$, under of the action of the diffeomorphism

$$f: \ u = \alpha \bullet u', v = v' \ (\alpha > 0), \tag{6.44}$$

are **isometrically equivalent**.

Indeed, if we act on the form (6.4) with the diffeomorphism (6.44), then for the case $\overline{\rho}_{12} > 0$ we get:

$$(ds)^2 = R^2 \left[(du)^2 + \sin^2(u)(dv)^2 \right] = R^2 \left[(d(\alpha u'))^2 + \sin^2(\alpha u')(dv)^2 \right] =$$

$$= R^2 \left[\alpha^2 (du')^2 + \sin^2(\alpha u')(dv)^2 \right] = (ds')^2 \Rightarrow ds = ds'.$$

Conclusion. For the case $\bar{\rho}_{12} > 0$, for the metric forms (6.4) and (6.33), the transformations $f: u = \alpha u'$, $v = v'$, where $\alpha > 0$, are **isometric equivalence.**

III. Equiareality

According to the **Definition** 5.4, for the coefficients of metric forms (6.4) and (6.33) **equiareality,** under the action of the diffeomorphism (6.42), for the case $\bar{\rho}_{12} > 0$ means the fulfillment of the identity:

$$\frac{\partial(u,v)}{\partial(u',v')} \equiv \sqrt{\frac{E'G' - (F')^2}{EG - F^2}} \; . \tag{6.45}$$

We show now, that for our situation the standard spherical form (6.4) and parameterized metric form (6.33) for the case $\bar{\rho}_{12} > 0$, under the action of the diffeomorphism (6.44), are **equiareal.**

Indeed, if we act on the form (6.4) with the diffeomorphism (6.44), then for the case $\bar{\rho}_{12} > 0$ we get:

$$\frac{\partial u}{\partial u'} = \alpha > 0, \quad \frac{\partial u}{\partial v'} = 0, \quad \frac{\partial v}{\partial u'} = 0, \quad \frac{\partial v}{\partial v'} = 1, \quad \frac{\partial(u,v)}{\partial(u',v')} = \alpha > 0, \tag{6.46}$$

$$\alpha = \frac{\partial(u,v)}{\partial(u',v')} = \sqrt{\frac{E'G' - (F')^2}{EG - F^2}} = \sqrt{\frac{R^2\alpha^2 \sin^2(\alpha u' = u)}{R^2 \sin^2(u)}} = \alpha > 0. \tag{6.47}$$

Comparing the corresponding relations (6.46) and (6.47) for the case $\alpha > 0$, we get the identity (6.45).

Conclusion. For the case $\bar{\rho}_{12} > 0$ for the metric forms (6.4) and (6.33), the transformations $f: u = \alpha u'$, $v = v'0$, where $\alpha > 0$, are **equiareality** .

6.2.4.2. Individual metric properties of the parameterized spherical metric forms in comparison with the standard spherical metric form at $\bar{\rho}_{12} > 0$

I. Violation of the property of isometric identity when $\bar{\rho}_{12} > 0$

According to the **Definition** 5.2, for the coefficients of the metric forms (6.4) and (6.33) the **isometric identity,** under the action of diffeomorphism (6.44) for the case $\bar{\rho}_{12} > 0$ means the fulfillment of the identities:

$$E'(u',v') \equiv E(u,v), \; F'(u',v') \equiv F(u,v), \; G'(u',v') \equiv G(u,v), \tag{6.48}$$

where $f: u = u(u',v') = \alpha \bullet u, \; v = v(u',v') = v..$

We show now that for this situation the isometric identity is not performed. Assume the contrary. Then, for the case $\alpha > 0$, $\overline{\rho}_{12} > 0$ we get the following identities:

$$E'(u',v') = R^2\alpha^2 \equiv E(u,v) = R^2 \Rightarrow \alpha = 1 \qquad (6.49)$$

$$F'(u',v') = 0 \equiv F(u,v) = 0 \Rightarrow 0 = 0, \qquad (6.50)$$

$$G'(u',v') = R^2 \sin^2(\alpha u') \equiv G(u,v) = R^2 \sin^2(u) \qquad (6.51)$$

Because the diffeomorphism (6.44) has the form $f: u = \alpha u', v = v', \alpha > 0$, then the formula (6.51) can be rewritten as follows:

$$G'(u',v') = R^2 \sin^2(u) \equiv G(u,v) = R^2 \sin^2(u) \Rightarrow 1 = 1 \qquad (6.52)$$

Thus, by using (6.49) and (6.52), we get that $\alpha = 1 \Rightarrow \overline{\rho}_{12} = 0$, what is impossible, because we have assumed that, under the action of the diffeomorphism (6.44), the form (6.4) is converted into the form (6.33), for which $\overline{\rho}_{12} > 0$.

Conclusion. For the case $\overline{\rho}_{12} > 0$, for the metric forms (6.4) and (6.33), the transformations $f : u = \alpha u', v = v'0$, where $\alpha > 0$, are **not isometric identity.**

II. **Violation of the property of conformality at** $\overline{\rho}_{12} > 0$

According to the **Definition** 5.3, for the coefficients of the metric forms (6.4) and (6.33) the **conformality**, under the action of the diffeomorphism (6.44) at $\overline{\rho}_{12} > 0$, means the fulfillment of the following identities:

$$\frac{E'(u',v')}{E(u,v)} \equiv \frac{F'(u',v')}{F(u,v)} \equiv \frac{G(u',v')}{G(u,v)} \equiv m^2 > 0, \qquad (6.53)$$

where $m = m(u,v) > 0$ is the **coefficient of conformality**.

We show now that for this situation the isometric conformality is not performed. Assume the contrary. Then for the case $\alpha > 0$, $\overline{\rho}_{12} > 0$ we get the identities:

$$E'(u',v') = R^2\alpha^2 \equiv m^2 E(u,v) = m^2 R^2 \Rightarrow \alpha = m > 0 \qquad (6.54)$$

$$F'(u',v') = 0 \equiv m^2 F(u,v) = m^2 \bullet 0 = 0, \qquad (6.55)$$

$$G'(u',v') = R^2 \sin^2(\alpha u') \equiv m^2 G(u,v) = m^2 R^2 \sin^2(u) \Rightarrow |\sin(\alpha u')| \equiv m|\sin(u)| > 0 \qquad (6.56)$$

where $u \neq \pi k, k \in Z$.

Substituting the value $\alpha = m > 0$ into (6.56) (see the condition (6.54), we get the following identity:

$$\left|\sin(mu')\right| \equiv m\left|\sin(u)\right| \Rightarrow m \equiv \left|\frac{\sin(mu')}{\sin(u)}\right|. \tag{6.57}$$

Since $\alpha = m > 0$, f: $u = \alpha \bullet u'$, $v = v'$, then the diffeomorphism has the following form:

$$f: u = mu', v = v'. \tag{6.58}$$

Substituting (6.58) into (6.57), we get $m \equiv \left|\frac{\sin(mu')}{\sin(u)}\right| = \left|\frac{\sin(u)}{\sin(u)}\right| = 1$, it follows

from here: $\alpha = m = 1$, $\overline{\rho}_{12}$ $\overline{\rho} = \frac{\rho_{12}}{R} = \left|\alpha - 1\right| = \left|1 - 1\right| = 0$, what is not possible, because in this case the condition $\overline{\rho}_{12} > 0$ should be fulfilled.

Conclusion. For the case $\overline{\rho}_{12} > 0$, for the metric forms (6.4) and (6.33), the transformations $f : u = \alpha u'$, $v = v'0$, where $\alpha > 0$, are **not conformal**.

6.3. The Spherical Mathematics of Harmony

6.3.1. Interpretation of the parameterized spherical forms in the terms of the metallic proportions

Recall that the Argentinean mathematician **Vera W. de Spinadel** has been introduced in [52] a special class of the mathematical constants, called the **metallic means**. The *metallic means* or *metallic proportions* are real numbers Φ_λ given by the following formula:

$$\Phi_\lambda = \frac{\lambda + \sqrt{4 + \lambda^2}}{2}, \tag{6.59}$$

where we assume that $\lambda \neq 0$ is a given real number.

In the works [13,14], the mathematical constants (6.59) have been used for the introduction of the new class of hyperbolic functions, the called **hyperbolic Fibonacci λ-functions.** Taking into consideration this fact, we will call the mathematical constants (6.59) the **hyperbolic metallic proportions.**

Beside of the function Φ_λ (6.59), we need the function $\ln\Phi_\lambda$. The graphs of these functions are presented in Fig.6.3.

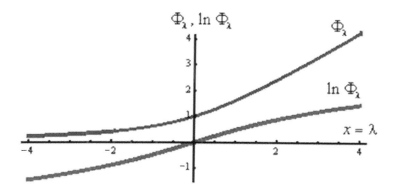

Figure 6.3. The graphs of the functions Φ_λ *and* $\ln \Phi_\lambda$

Let us introduce now a very important mathematical concept.

Definition 6.1. *Spherical metallic proportion* *is called a complex number of the following form:*

$$\tau_\lambda = \Phi_\lambda^i = \cos(\ln \Phi_\lambda) + i \sin(\ln \Phi_\lambda), \quad \lambda \neq 0, \quad i = \sqrt{-1}. \tag{6.60}$$

Note that because $\operatorname{Re}(\tau_\lambda) = \cos(\ln \Phi_\lambda)$, $\operatorname{Im}(\tau_\lambda) = \sin(\ln \Phi_\lambda)$, then the ***spherical metallic proportion*** can also be interpreted as a pair of real numbers, which are **real** and **imaginary** parts of the complex number (6.60). We write this complex number as follows:

$$[\cos(\ln \Phi_\lambda), \sin(\ln \Phi_\lambda)], \lambda \neq 0 \tag{6.61}$$

The graphs of the functions $\operatorname{Re}(\tau_\lambda) = \cos(\ln \Phi_\lambda)$ and $\operatorname{Im}(\tau_\lambda) = \sin(\ln \Phi_\lambda)$ are presented in Fig. 6.4 and 6.5.

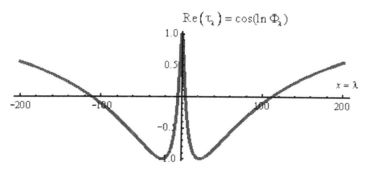

Figure 6.4. The graph of the function $\operatorname{Re}(\tau_\lambda) = \cos(\ln \Phi_\lambda)$

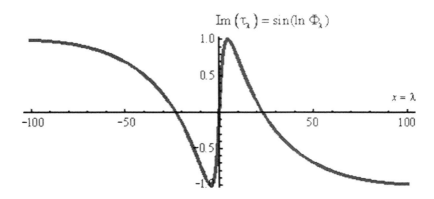

$$\text{Im}\left(\tau_{\lambda}\right)=\sin(\ln\Phi_{\lambda})$$

Figure 6.5. The graph of the function $\text{Im}(\tau_{\lambda})=\sin(\ln\Phi_{\lambda})$

Definition 6.2. *We call the spherical metallic proportions (6.61) by the* **golden, silver, bronze, copper proportions** *for the case* $\lambda=1,2,3,4$ *in (6.61), respectively.*

We introduce the designation $\tau_{1}=\tau$. By using (6.59) and (6.61), at $\lambda=1,2,3,4$ we get the following numerical values for the above *spherical metallic proportions* (6.61):

The **golden** *spherical proportion:* $\tau=[\cos(\ln\Phi),\sin(\ln\Phi)]\approx(0.8864,0.4628)$

The **silver** *spherical proportion:* $\tau_{2}=[\cos(\ln\Phi_{2}),\sin(\ln\Phi_{2})]\approx(0.6969,0.07171)$

The **bronze** *spherical proportion:* $\tau_{3}=[\cos(\ln\Phi_{3}),\sin(\ln\Phi_{3})]\approx(0.5081,0.8612)$

The **copper** *spherical proportion:* $\tau_{4}=[\cos(\ln\Phi_{4}),\sin(\ln\Phi_{4})]\approx(0.3361,0.9418)$

Definition 6.3. *We call* **spherical λ-sine** *and* **spherical λ-cosine** *the following functions, respectively:*

$$SF_{\lambda}(x)=-i\frac{\tau_{\lambda}^{x}-\tau_{\lambda}^{-x}}{\sqrt{4+\lambda^{2}}}=\frac{2}{\sqrt{4+\lambda^{2}}}\sin\left(x\ln\Phi_{\lambda}\right) \qquad (6.62)$$

$$CF_{\lambda}(x)=\frac{\tau_{\lambda}^{x}+\tau_{\lambda}^{-x}}{\sqrt{4+\lambda^{2}}}=\frac{2}{\sqrt{4+\lambda^{2}}}\cos\left(x\ln\Phi_{\lambda}\right) \qquad (6.63)$$

where the symbols S and C mean the spherical λ-sine and λ-cosine, respectively.

We rewrite the formulas (6.62) and (6.63) as follows:

$$SF_\lambda(x) = \frac{2}{\sqrt{4+\lambda^2}}\sin(y) \qquad (6.64)$$

$$CF_\lambda(x) = \frac{2}{\sqrt{4+\lambda^2}}\cos(y), \qquad (6.65)$$

where $y = x\ln\Phi_\lambda = \ln\Phi_\lambda^x$.

It follows from here:

$$\sin(y) = \frac{\sqrt{4+\lambda^2}}{2}\times\left[SF_\lambda(x)\right], \quad \cos(y) = \frac{\sqrt{4+\lambda^2}}{2}\times\left[CF_\lambda(x)\right]..$$

From the properties of $\sin(y), \cos(y)$, there follow the relevant properties for the spherical Fibonacci λ-functions $SF_\lambda(x), CF_\lambda(x)$. Consider any of these properties in comparison to the corresponding properties of the classic trigonometric functions $\sin(y), \cos(y)$:

$$\begin{cases} \sin^2(y) + \cos^2(y) = 1 \Leftrightarrow \left[SF_\lambda(x)\right]^2 + \left[CF_\lambda(x)\right]^2 = \dfrac{4}{4+\lambda^2}, \\[2mm] \sin(y_1 \pm y_2) = \sin(y_1)\cos(y_2) \pm \cos(y_1)\sin(y_2) \Leftrightarrow \\[1mm] \Leftrightarrow SF_\lambda(x_1 \pm x_2) = \dfrac{\sqrt{4+\lambda^2}}{2}\left[SF_\lambda(x_1)CF_\lambda(x_2) \pm CF_\lambda(x_1)SF_\lambda(x_2)\right] \\[2mm] \cos(y_1 \pm y_2) = \cos(y_1)\cos(y_2) \mp \sin(y_1)\sin(y_2) \Leftrightarrow \\[1mm] \Leftrightarrow SF_\lambda(x_1 \pm x_2) = \dfrac{\sqrt{4+\lambda^2}}{2}\left[CF_\lambda(x_1)CF_\lambda(x_2) \mp SF_\lambda(x_1)SF_\lambda(x_2)\right] \end{cases}$$

and so on

We compare now the *spherical Fibonacci λ-functions* (6.62) and (6.63) (the *spherical Fibonacci λ-sine* $SF_\lambda(x)$ and the *spherical Fibonacci λ-cosine* $CF_\lambda(x)$ with the corresponding *hyperbolic Fibonacci λ-functions* (the *hyperbolic Fibonacci λ-sine* $sF_\lambda(x)$ and the *hyperbolic λ-cosine* $cF_\lambda(x)$).

As is known [13,14], the *hyperbolic Fibonacci λ-sine* $sF_\lambda(x)$ and the *hyperbolic Fibonacci λ-cosine* $cF_\lambda(x)$ are given by the following formulas:

$$sF_\lambda(x) = \frac{\Phi_\lambda^x - \Phi_\lambda^{-x}}{2} = \frac{2}{\sqrt{4+\lambda^2}}sh(x\ln\Phi_\lambda) \qquad (6.66)$$

$$cF_\lambda(x) = \frac{\Phi_\lambda^x + \Phi_\lambda^{-x}}{2} = \frac{2}{\sqrt{4+\lambda^2}}ch(x\ln\Phi_\lambda) \qquad (6.67)$$

It follows from here the following relations between the *spherical Fibonacci λ-functions* and the *hyperbolic Fibonacci λ-functions*:

$$SF_\lambda(x) = \frac{\sin(x \ln \Phi_\lambda)}{sh(x \ln \Phi_\lambda)} sF_\lambda(x), \quad CF_\lambda(x) = \frac{\cos(x \ln \Phi_\lambda)}{ch(x \ln \Phi_\lambda)} cF_\lambda(x) \qquad (6.68)$$

$$sF_\lambda(x) = \frac{sh(x \ln \Phi_\lambda)}{\sin(x \ln \Phi_\lambda)} SF_\lambda(x), \quad cF_\lambda(x) = \frac{ch(x \ln \Phi_\lambda)}{\cos(x \ln \Phi_\lambda)} CF_\lambda(x) \qquad (6.69)$$

The formulas (6.68) and (6.69) allow us to find all the relationships for the spherical "mathematics of harmony" by using hyperbolic Fibonacci λ-functions. We demonstrate below the examples of obtaining the various formulas of the spherical "mathematics of harmony" based on hyperbolic Fibonacci λ-functions.

We make now the following replacement $y = x \ln \Phi_\lambda = \ln \Phi_\lambda^x$ in the relations (6.66), (6.67). Then, we get:

$$sF_\lambda(x) = \frac{2}{\sqrt{4+\lambda^2}} sh(y); \quad cF_\lambda(x) = \frac{2}{\sqrt{4+\lambda^2}} ch(y)$$

$$sh(y) = \frac{\sqrt{4+\lambda^2}}{2} sF_\lambda(x); \quad ch(y) = \frac{\sqrt{4+\lambda^2}}{2} cF_\lambda(x)$$

From the properties of the classical hyperbolic functions $sh(y), ch(y)$, there follow the corresponding relations for the hyperbolic Fibonacci λ-functions $sF_\lambda(x), cF_\lambda(x)$. Let us consider some of these relations:

$$ch^2(y) - sh^2(y) = 1 \Leftrightarrow \left[cF_\lambda(x)\right]^2 - \left[sF_\lambda(x)\right]^2 = \frac{4}{4+\lambda^2},$$

$$sh(y_1 \pm y_2) = sh(y_1)ch(y_2) \pm ch(y_1)sh(y_2) \Leftrightarrow$$

$$\Leftrightarrow sF_\lambda(x_1 \pm x_2) = \frac{\sqrt{4+\lambda^2}}{2} [sF_\lambda(x_1)cF_\lambda(x_2) \pm cF_\lambda(x_1)sF_\lambda(x_2)],$$

$$ch(y_1 \pm y_2) = ch(y_1)ch(y_2) \pm sh(y_1)sh(y_2) \Leftrightarrow$$

$$\Leftrightarrow cF_\lambda(x_1 \pm x_2) = \frac{\sqrt{4+\lambda^2}}{2} [cF_\lambda(x_1)cF_\lambda(x_2) \pm cF_\lambda(x_1)cF_\lambda(x_2)]$$

and so on.

At the replacement $\alpha = \ln \Phi_\lambda > 0$ ($\lambda > 0$), the parameterized spherical metric form (6.33) can be rewritten as follows:

$$(ds')^2 = R^2 \left[(\ln^2 \Phi_\lambda)(du')^2 + \frac{4+\lambda^2}{4}(SF_\lambda(u'))^2 (dv')^2 \right] \qquad (6.70)$$

where $SF_\lambda(u')$ is the spherical Fibonacci λ-sine, $\dfrac{k}{\ln \Phi_\lambda}\pi < u' < \dfrac{k+1}{\ln \Phi_\lambda}\pi, -\infty < v' < +\infty$,

$k \in Z$ (Z is a set of integers).

We call the form (6.70) the **spherical metric λ-form.**

Because, according to (6.62), the spherical Fibonacci λ-sine has the form

$SF_\lambda(u') = \dfrac{2}{\sqrt{4+\lambda^2}}\sin\left(\ln \Phi_\lambda^{u'}\right)$, then the spherical metric λ-form (6.70) can be rewritten as follows:

$$(ds')^2 = R^2\left[\left(\ln^2 \Phi_\lambda\right)(du')^2 + \sin^2\left(\ln \Phi_\lambda^{u'}\right)(dv')^2\right] \tag{6.71}$$

At the value

$$\lambda = 2\mathrm{sh}(1) \approx 2.3504 \Rightarrow \Phi_\lambda = e \approx 2.7182 \Rightarrow \ln \Phi_\lambda = 1, \tag{6.72}$$

the spherical λ-form (6.71) coincides with the standard metrical form (6.4), here we assume: $u = u', v = v'$.

Similarly, because, according to (6.66), the hyperbolic Fibonacci λ-sine has the following form $sF_\lambda(u') = \dfrac{\Phi_\lambda^{u'} - \Phi_\lambda^{-u'}}{\sqrt{4+\lambda^2}} = \dfrac{2}{\sqrt{4+\lambda^2}}sh\left(\ln \Phi_\lambda^{u'}\right)$, then

Lobachevski's hyperbolic metric λ-form, introduced by the authors in the works [29-31]

$$(ds')^2 = R^2\left[\left(\ln^2 \Phi_\lambda\right)(du')^2 + \dfrac{4+\lambda^2}{4}\left(sF_\lambda(u')\right)^2(dv')^2\right] \tag{6.73}$$

can be rewritten as follows:

$$(ds')^2 = R^2\left[\left(\ln^2 \Phi_\lambda\right)(du')^2 + sh^2\left(\ln^2 \Phi_\lambda^{u'}\right)(dv')^2\right]. \tag{6.74}$$

At the value of the parameter $\lambda = 2\mathrm{sh}(1) \approx 2.3504$, *Lobachevski's hyperbolic metric λ-form* (6.73) coincides with the classical Lobachevski's metric form:

$$(ds)^2 = R^2\left[(du)^2 + sh^2(u)(dv)^2\right], \tag{6.75}$$

here we assume: $u \equiv u', v \equiv v'$.

6.3.2. The normalized distance between the spherical metric λ-forms and standard spherical metric form

For each fixed value $R>0$, we introduce the **distance** ρ_{12} between the spherical metric λ-forms (6.70), where $\dfrac{k}{\ln \Phi_\lambda}\pi < u' < \dfrac{k+1}{\ln \Phi_\lambda}\pi, -\infty < v' < +\infty, \ k \in Z$ $\lambda > 0$, and the standard spherical metric form (6.4), in the following form:

$$\rho_{12} = R\left|(\ln \Phi_\lambda)-1\right| . \tag{6.76}$$

In the future, unless otherwise stated, we will consider only the **normalized distance** $\overline{\rho}_{12}$ between the metric forms (6.70) and (6.4):

$$\overline{\rho}_{12} = \frac{\rho_{12}}{R} = \left|(\ln \Phi_\lambda)-1\right|, \ \lambda > 0 . \tag{6.77}$$

In other words, the normalized distance $\overline{\rho}_{12}$ in (6.76) is obtained from the **normalized distance** $\overline{\rho}_{12}$ in (6.38) by means of the replacement $\alpha = \ln \Phi_\lambda$. The graph of the normalized distance $\overline{\rho}_{12}$ in (6.77), depending on the parameter $\lambda > 0$, is presented in Fig. 6.6.

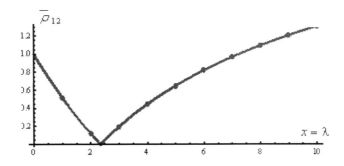

Figure 6.6. The graph of the function $\overline{\rho}_{12}$ at $\lambda>0$.

At $\overline{\rho}_{12}=0 \Leftrightarrow \Phi_\lambda = e \approx 2.7182$, where $\lambda = 2\mathrm{sh}(1) \approx 2.3504$, the spherical metric λ-form (6.70) of the sphere S^2 with the radius R coincides with the standard metric form (6.4) of the sphere S^2 with the radius R for the identity mapping $u \equiv u', v \equiv v'$.

Table 6.1 gives the numerical values of the normalized distance $\overline{\rho}_{12}$ depending on the integer values of the coefficient $\lambda>0$.

Table 6.1. The numerical values of the normalized distance $\bar{\rho}_{12}$

λ	Name of metallic proportions	Φ_λ	Approximate values of Φ_λ	The normalized distance $\bar{\rho}_{12}$
1	Golden	$\dfrac{1+\sqrt{5}}{2}$	1.618	0.518
2	Silver	$\dfrac{1+\sqrt{2}}{~}$	2.414	0.118
3	Bronze	$\dfrac{3+\sqrt{13}}{2}$	3.303	0.1947
4	Cooper	$2+\sqrt{5}$	4.236	0.443
5	–	$\dfrac{5+\sqrt{29}}{2}$	5.192	0.6472
6	–	$3+\sqrt{10}$	6.162	0.8184

There is the following relation between the metallic proportion Φ_λ and its reciprocal $\dfrac{1}{\Phi_\lambda}$:

$$\Phi_\lambda = \lambda + \frac{1}{\Phi_\lambda} \tag{6.78}$$

It follows from (6.78) that for the case $\lambda = 1,2,3,\ldots$ the metallic proportion Φ_λ has a unique property. It follows from Table 6.1 that for the case $\lambda = 1,2,3,\ldots$, every metallic proportion Φ_λ is a mixed fraction, which consists of integer part and mantissa. According to (6.78), the integer part of the mixed fraction Φ_λ is equal to the parameter $\lambda = 1,2,3,\ldots$ and the mantissa m_λ is equal to the reciprocal $\dfrac{1}{\Phi_\lambda}$, that is $m_\lambda = \dfrac{1}{\Phi_\lambda}$. Table 6.2 demonstrates this unique property called the *property of mantissa preservation*.

Table 6.2. Property of mantissa preservation

λ	1	2	3	4	5	6
Φ_λ	1.618	2.414	3.303	4.236	5.192	6.162
$m_\lambda = \dfrac{1}{\Phi_\lambda}$	0.618	0.414	0.303	0.236	0.192	0.162

Conclusions

1. The minimal distance. Comparing the *metallic proportions* according to the *normalized distance* $\bar{\rho}_{12}$ (Table 6.1), we can conclude, that the *silver proportion* $1+\sqrt{2} \approx 2.414$ ($\lambda = 2$) has the **minimal distance** $\bar{\rho}_{12} = 0.118$ among all other metallic proportions Φ_λ for the positive integer values of the parameter $\lambda = 1,2,3,...$. This means that among the *spherical metric λ-forms*, generated by the metallic proportions Φ_λ for the integer values of $\lambda = 1,2,3,...$, the *"silver"* *spherical metric λ-form*, generated by the *silver proportion* $1+\sqrt{2} \approx 2.414$ ($\lambda = 2$), is the **closest spherical metric λ-form** to the standard metric spherical form (6.4), in the sense of the normalized distance $\bar{\rho}_{12}$.

2. Property of mantissa preservation: for the integer $\lambda = 1,2,3,...$, the mantissa m_λ of the mixed fraction Φ_λ coincides with its reciprocal $\dfrac{1}{\Phi_\lambda}$, that is, $m_\lambda = \dfrac{1}{\Phi_\lambda}$.

6.4. The solution of Hilbert's Fourth Problem on the basis of the Mathematics of Harmony

6.4.1. One more excursus to Hilbert's Problems

6.4.1.1. Hilbert's Fourth Problem

The 23 mathematical problems, presented by **David Hilbert** (1862-1943) at the II International Congress of Mathematicians (Paris, August 6 - 12, 1900) are widely known in mathematics as the title *Hilbert's Problems*. In his work "Mathematical Problems" [6] Hilbert formulated the fundamental mathematical problems, which, in his opinion, are the most significant for the further development of many different branches of mathematics. For example, such as the *foundations of mathematics, algebra, number theory, geometry, topology, algebraic geometry, Lie groups, real and complex analysis, differential equations, mathematical physics and probability theory, calculus of variations.*

According to [72], the sixteen of twenty-third problems are resolved. One more two problems are not correct mathematical problems because **Hilbert's Fourth Problem** was formulated too vague [8] and **Hilbert's Sixth Problem** is more physical rather than mathematical and therefore far from solution. Among the five remaining problems, the two of them (**Hilbert's Eighth and Twelfth Problems**) are not resolved, and three of them (**Hilbert's Ninth, Fifteenth and Sixteenth Problems**) have been resolved only for certain cases.

In Chapter 6 we turn our attention to **Hilbert's Fourth Problem** and consider one of the possible solutions of this problem for the case of spherical geometry.

In Hilbert's formulation [6], the Fourth Problem is geometric problem and is called *"Problem of the straight line as the shortest distance between two points."*

The preamble to Hilbert's work [6] is crucial for understanding of Hilbert's philosophy and his approach to "Hilbert's Problems." It is important to emphasize that even in the preamble **Gilbert emphasizes the particular importance of the Fourth Problem for the development of mathematics**. He says:

"But it often happens also that the same special problem finds application in the most unlike branches of mathematical knowledge. So, for example, the problem of the shortest line plays a chief and historically important part in the foundations of geometry, in the theory of curved lines and surfaces, in mechanics and in the calculus of variations. And how convincingly has F. Klein, in his work on the icosahedron, pictured the significance which attaches to the problem of the regular polyhedra in elementary geometry, in group theory, in the theory of equations and in that of linear differential equations."

Note that this quote contains the mention about very interesting work, which has the same importance for the development of mathematics as all "Hilbert's Problems." We are talking about the book of the famous German mathematician **Felix Klein "Lectures on Icosahedron"** [41]. Below we will make more detailed analysis of the influence of Klein's "icosahedral idea" on the development of mathematics.

In the section, concerning the Fourth Problem, Hilbert gives a more specific statement of the problem:

"The more general question now arises: Whether from other suggestive standpoints geometries may not be devised which, with equal right, stand next to Euclidean geometry."

Hilbert analyzes the geometries, which, on his opinion, *"stand next to Euclidean geometry"*:

*"Another problem relating to the foundations of geometry is this: If from among the axioms necessary to establish ordinary Euclidean geometry, we exclude the axiom of parallels, or assume it as not satisfied, but retain all other axioms, we obtain, as is well known, **the geometry of Lobachevski (hyperbolic geometry)**. We may therefore say that this is a geometry standing next to Euclidean geometry."*

Besides hyperbolic geometry, Hilbert also considers "*Riemann's geometry* or *elliptic geometry*. He emphasizes *"that this geometry appears to be the next after Lobachevski's geometry."*

Then, Hilbert emphasizes that **Minkowski's geometry** *"is therefore also a geometry standing next to the ordinary Euclidean geometry."*

In other words, according to Hilbert, the three known for Hilbert geometries, **Lobachevski's geometry, Riemann's geometry** and **Minkovski's geometry** are the closest to the Euclidean geometry.

Note, that we should distinguish **Riemann's geometry** from the **Riemannian geometry** [4]. **Riemann's geometry** (*elliptic geometry*) is the geometry of two-dimensional sphere with identified diametrically opposite points, that is, the geometry of the *projective plane*. Hilbert pays a special attention to **Riemann's geometry**, because on the projective plane any two geodesic lines always overlap, but they have not two, as in the two-dimensional sphere, but only one point of intersection.

Hilbert considers Lobachevski's geometry and Riemann's geometry *"in some sense"* the closest to the Euclidean geometry. Similarly, *"in some sense"* the geometry of two-dimensional sphere can be considered the closest to the Euclidean geometry, because the geometry of two-dimensional sphere is the *universal covering* for the geometry of projective plane and both geometries have the same constant positive Gaussian curvature.

The **Riemannian geometry** studies smooth multidimensional manifolds with additional structure – the *Riemannian metrics*. The **Riemannian geometry** is a generalization of the internal geometry for two-dimensional surfaces and contains it as its constituent part.

However, a brilliant scientific Hilbert's intuition tells him that these geometries are not the only geometries, which are close to Euclidean geometry, and he puts the fundamental problem of searching new non-Euclidean geometries, **standing next to Euclidean geometry."**

In conclusion of this section, he says:

"The theorem of the straight line as the shortest distance between two points and the essentially equivalent theorem of Euclid about the sides of a triangle, play an important part not only in number theory but also in the theory of surfaces and in the calculus of variations. For this reason, and because I believe that the thorough investigation of the conditions for the validity of this theorem will throw a new light upon the idea of distance, as well as upon other elementary ideas, e. g., upon the idea of the plane, and the possibility of its definition by means of the idea of the straight line, the construction and systematic treatment of the geometries here possible seem to me desirable".

6.4.1.2. Hilbert's methodology and philosophy

Hilbert's work [6] is of great interest from philosophical and methodological point of view. Hilbert's arguments, concerning various aspects of solving mathematical problems, are of valuable importance. In the preamble to "Hilbert's Problems" [6] he writes:

"History teaches the continuity of the development of science. We know that every age has its own problems, which the following age either solves or casts aside as profitless and replaces by new ones. If we would obtain an idea of the probable development of mathematical knowledge in the immediate future, we must let the unsettled questions pass before our minds and look over the problems which the science of today sets and whose solution we expect from the future. To such a review of problems the present day, lying at the meeting of the centuries, seems to me well adapted. For the close of a great epoch not only invites us to look back into the past but also directs our thoughts to the unknown future.

The deep significance of certain problems for the advance of mathematical science in general and the important role which they play in the work of the individual investigator are not to be denied. As long as a branch of science offers an abundance of problems, so long is it alive; a lack of problems foreshadows extinction or the cessation of independent development. Just as every human undertaking pursues certain objects, so also mathematical research requires its problems. It is by the solution of problems that the investigator tests the temper of his steel; he finds new methods and new outlooks, and gains a wider and freer horizon."

Hilbert warns about the dangers of modern isolation of "pure mathematics" from the experience:

"In the meantime, while the creative power of pure reason is at work, the outer world again comes into play, forces upon us new questions from actual experience, opens up new branches of mathematics, and while we seek to conquer these new fields of knowledge for the realm of pure thought, we often find the answers to old unsolved problems and thus at the same time advance most successfully the old theories. **And it seems to me that the numerous and surprising analogies and that apparently prearranged harmony which the mathematician so often perceives in the questions, methods and ideas of the various branches of his science, have their origin in this ever-recurring interplay between thought and experience**.

This wonderful quotation contains some important thoughts. The first one consists in the fact that *"while the creative power of pure reason is at work, the outer world again comes into play."* By using experience, *"we often find the answers to old unsolved problems and thus at the same time advance most successfully the old theories."* The second important thought is that *"**the numerous and surprising analogies and that apparently prearranged harmony ... have their origin in this ever-recurring interplay between thought and experience**.*

In this quote Hilbert refers to the idea of *"**pre-arranged harmony**"* (or *Leibniz's Doctrine about Pre-established Harmony*), which dates back in its origin to the *"**ever-recurring interplay between thought and experience**."*

As general requirements for the solution of mathematical problems, Hilbert puts forth **rigor** and **simplicity**, while he emphasizes:

"Besides it is an error to believe that rigor in the proof is the enemy of simplicity. On the contrary we find it confirmed by numerous examples that the **rigorous method is at the same time the simpler and the more easily comprehended.** *The very effort for rigor forces us to find out simpler methods of proof. It also frequently leads the way to methods which are more capable of development than the old methods of less rigor.*

Then Hilbert expresses the idea that the **strict simple proof** can be implemented as on the language of **mathematical formulas**, and in the **geometric form**, because *"the arithmetical symbols are written diagrams and the geometrical figures are graphic formulas."*

Applying these valuable ideas specifically to the solution of **Hilbert's Fourth Problem**, which is **geometric problem**, we can draw the following conclusion.

The "*game between thought and experience,*" at the vision of **Hilbert's Fourth Problem** as **vague** [9,72], means the **game** between the **axiomatic approach** and the language of the **strict and simple formulas of the Mathematics of Harmony** [19], which has not only scientific but also great practical importance for the natural sciences. In [76] the similar idea is expressed as **the game between "postulates" and "hyperbolic functions."**

It should be noted that the axiomatic approach to solving Hilbert's Fourth Problem was used in the second half of 20 c. by the famous Russian and Ukrainian mathematician **A.V. Pogorelov** (1919-2002) [71]. However, as mentioned, at present the majority of mathematicians recognized that Hilbert's Fourth Problem is too vague to understand solved it or not [8],[72]. Therefore, it is not correct to insist that Pogorelov completely solved this problem. As pointed out by Aranson [73], apparently, we are talking about a partial solution of the problem based on the axiomatic approach.

But there is another approach to solving this problem, for which we also can not argue that this complex and important mathematical problem is completely solved. This approach to solving Hilbert's Fourth Problem consists in the use of the new classes of hyperbolic functions, which arose in the "mathematics of harmony" [19].

6.4.2. One more about Klein's "icosahedral idea"

6.4.2.1. The "icosahedral idea" by Felix Klein and Proclus' hypothesis

And now we turn again to the important quote from Hilbert's work [6] regarding Felix Klein's book [41]:

"*And how convincingly has F. Klein, in his work on the icosahedron, pictured the significance which attaches to the problem of the regular polyhedra in elementary geometry, in group theory, in the theory of equations and in that of linear differential equations.*"

The name of the German mathematician **Felix Klein** (1849 –1925) is well known in mathematics. In the 19th century **Felix Klein** tried to unite all branches of mathematics on the base of the *regular icosahedron* dual to the *dodecahedron* [41].

Like "Hilbert's Problems," Klein's "icosahedral idea" is very important for the development of mathematics. Klein interprets the *regular icosahedron*, based on the *golden ratio,* as a geometric object, which connects the 5 mathematical theories: *geometry, Galois theory, group theory, invariant theory,* and *differential equations.* Klein's main idea is extremely simple: *"Each unique geometric object is connected one way or another with the properties of the regular icosahedron."* Unfortunately, this remarkable idea was not developed in contemporary mathematics, which is one of the **"strategic mistakes"** in the development of mathematics [18].

As pointed out by Hilbert, Klein's "icosahedral idea" in its origins is dating back to the **regular polyhedra** or **Platonic solids**, which are associated in ancient Greece with the "Harmony of the Universe" [26].

It is well known that the Book XIII, that is, the final book of Euclid's *Elements* is devoted to the geometric theory of *Platonic solids.* The Greek philosopher and mathematician Proclus was the first, who drew attention to this fact. From this fact Proclus made the surprising conclusion that the main purpose of Euclid at the creation of the *Elements* was to give a complete geometric theory of *Platonic Solids.* This means that Euclid's *Elements* can be considered as historically the first version of the "Harmony Mathematics," which, according to Proclus, was embodied in this outstanding mathematical work of Greek mathematics.

This approach leads to the conclusion, which can be unexpected for many mathematicians. It is found [18], that in parallel with the *Classical Mathematics,* another mathematical direction – the *Harmony Mathematics* – was studied in Greek science. Similarly to the *Classical Mathematics,* the *Harmony Mathematics* dates back in its origin to Euclid's *Elements.* However, the *Classical Mathematics* focuses on the *axiomatic approach,* while the *Harmony Mathematics* is based on the *golden ratio* (Proposition II.11) and *Platonic Solids* described in Book XIII of Euclid's *Elements.* Thus, Euclid's *Elements* is the source of two independent directions in the development of mathematics – the *Classical Mathematics* and the *Harmony Mathematics.*

6.4.2.3. The Classical Mathematics and the Mathematics of Harmony

For many centuries, the main attention of mathematicians was directed towards the creation of the *Classical Mathematics*, which became the *Tsarina of Natural Sciences*. However, the forces of many prominent thinkers and mathematicians, starting from **Pythagoras, Plato, Euclid, Pacioli, Kepler** up to **Lucas, Binet, Vorobyov, Hoggatt** and so forth, were directed towards the development of the basic concepts and applications of the *Harmony Mathematics*. Unfortunately, these important mathematical directions developed separately from one other. The time has come to unite them. This unusual union can lead to new scientific discoveries in mathematics and natural sciences. Some of the latest discoveries in the natural sciences, in particular, *Shechtman's quasi-crystals* based on Plato's icosahedron and *fullerenes* (Nobel Prize of 1996) based on the Archimedean truncated icosahedron require this union. All mathematical theories should be joined for one unique purpose to discover and explain Nature's Laws.

The modern "mathematics of harmony" [18,19] is the reflection and development of the ancient "Harmony Idea," embodied in Euclid's *Elements*. In the "Mathematics of Harmony," new and original mathematical results have been obtained, in particular, the **Fibonacci hyperbolic functions** [10-12], based Fibonacci numbers and "golden proportion," and their generalization, the **hyperbolic Fibonacci λ-functions** [13,14], based on the so-called λ-Fibonacci numbers and Spinadel's "metallic proportions" [52]. These classes of hyperbolic functions were used by the authors to obtain the original solution of Hilbert's Fourth Problem, set out in [29-32]. According to this solution, there is an infinite number of hyperbolic geometries, which are close to Lobachevski's classical geometry.

6.4.3. A solution of Hilbert's Fourth Problem on the basis of spherical and hyperbolic Fibonacci λ-functions

Previously, the authors of this article have published the works [29-32], where the authors suggested the original solution of Hilbert's Fourth Problem for the case of hyperbolic geometry. This solution includes an **endless set of the hyperbolic geometries**, induced by the **hyperbolic metric λ- forms of one and the same constant negative Gaussian curvature.**

In this chapter the authors suggests the original solution of Hilbert's Fourth Problem for the case of spherical geometry. This solution includes an **endless set of the spherical geometries**, induced by the **spherical metric λ- forms of one and the same constant positive Gaussian curvature.**

Both of these solutions are based on the use of the "strict and simple" formulas of the **"mathematics of harmony"** [19], which has not only theoretical interest, but also a wide application in natural sciences.

These two solutions of Hilbert's Fourth Problem completely are correlated with Hilbert's recommendations in finding the solutions for Hilbert's Problems, which are formulated not very clearly. According to Hilbert's recommendations, these solutions, on the one hand, should be **strict** and **simple** (the *"game with thinking"*), and on the other hand, should have, if possible, a **practical significance** (*the "game with experience"*).

Since the solution of Hilbert's Fourth Problem with the use of the **spherical metric λ-forms** is reduced to the consideration of the induced metric properties of **the two-dimensional sphere** in the three-dimensional space with the Euclidean metrics, then such solution of Hilbert's Fourth Problem will be called **spherical solution.**

Similarly, because the solution of Hilbert's Fourth Problem with the use of **Lobachevski's hyperbolic metric λ-forms** naturally is reduced to the consideration of the induced metric properties of the **pseudo-sphere** (the upper half of two-sheeted hyperboloid) in three-dimensional space with **Minkowski's alternating metrics**, then such solution of Hilbert's Fourth Problem will be called **pseudo-spherical solution.**

Below we present a comparative table of the spherical and pseudo-spherical solutions of Hilbert's Fourth Problem on the basis of the "Mathematics of Harmony" [19].

Table 6.3 (Steps 1-5). Comparison of the spherical and pseudo-spherical solutions of Hilbert's Fourth Problem

Step 1

Spherical solution	Pseudo-spherical solution
Basic surfaces	
Sphere $X^2 + Y^2 + Z^2 = R^2$ in the space (X,Y,Z) with the **Euclidean metrics** $(dl)^2 = (dX)^2 + (dY)^2 + (dZ)^2$. Here dl is an element of arc length in the space (X,Y,Z), $R > 0$ is the **radius** of the sphere, **Gaussian curvative** of the sphere is equal to $K = \dfrac{1}{R^2} > 0$.	**Pseudosphere,** the upper half of the two-sheeted hyperboloid $Z^2 - X^2 - Y^2 = R^2 > 0$, $Z > 0$ in the space (X,Y,Z) with Minkovski's metrics $(dl)^2 = (dZ)^2 - (dX)^2 - (dY)^2$ Here dl is an element of arc length in the space (X,Y,Z), $R > 0$ is the **radius** of the pseudosphere, **Gaussian curvative** of the pseudosphere is equal to $K = -\dfrac{1}{R^2} < 0$.
Parametric form of the basic surfaces	
$X = R\sin(u)\cos(v)$ $Y = R\sin(u)\sin(v)$ $Z = R\cos(u)$	$X = Rsh(u)\cos(v)$ $Y = Rsh(u)\sin(v)$ $X = Rch(u)$
Metric basic forms	
Standard spherical metric form of the Gaussian curvature $K = \dfrac{1}{R^2} > 0$ $(ds)^2 = R^2[(du)^2 + \sin^2(u)(dv)^2]$, $\pi k < u < \pi(k+1), -\infty < v < +\infty \ (k \in \mathbf{Z})$, ds is an element of arc length	**Lobachevski's metric form of the Gaussian curvature** $K = -\dfrac{1}{R^2} < 0$ $(ds)^2 = R^2[(du)^2 + sh^2(u)(dv)^2]$, $0 < u < +\infty, -\infty < v < +\infty$, ds is an element of arc length

Step 2

Spherical solution	Pseudo-spherical solution

Metallic proportions

Spherical proportions: $$\tau_\lambda = \Phi_\lambda^i = \cos(\ln\Phi_\lambda) + i\sin(\ln\Phi_\lambda),\ \lambda \neq 0,\ i = \sqrt{-1}$$	Hyperbolic proportions: $$\Phi_\lambda = \frac{\lambda + \sqrt{4+\lambda^2}}{2} = e^{\ln\Phi_\lambda},\ \lambda \neq 0$$

Partial cases of the proportions τ_λ and Φ_λ

$\lambda=1$: the golden spherical proportion $\tau_1 = \Phi_1^i$, $\lambda=2$: the silver spherical proportion $\tau_2 = \Phi_2^i$, $\lambda=3$: the bronze spherical proportion $\tau_3 = \Phi_3^i$, $\lambda=4$: the cooper spherical proportion $\tau_4 = \Phi_4^i$, $\lambda = \lambda^* = 2sh(1) \Rightarrow \tau_{\lambda^*} = e^i$	$\lambda=1$: the golden hyperbolic proportion Φ_1, $\lambda=2$: the silver hyperbolic proportion Φ_2, $\lambda=3$: the bronze hyperbolic proportion Φ_3, $\lambda=4$: the cooper hyperbolic proportion Φ_4, $\lambda = \lambda^* = 2sh(1) \Rightarrow \Phi_{\lambda^*} = e$

λ-sines and λ-cosines

Spherical λ-sine: $$SF_\lambda(x) = -i\frac{\tau_\lambda^x - \tau_\lambda^{-x}}{\sqrt{4+\lambda^2}} = \frac{2}{\sqrt{4+\lambda^2}}\sin\left(x\ln\Phi_\lambda\right)$$ Spherical $\lambda-$cosine: $$CF_\lambda(x) = \frac{\tau_\lambda^x + \tau_\lambda^{-x}}{\sqrt{4+\lambda^2}} = \frac{2}{\sqrt{4+\lambda^2}}\cos\left(x\ln\Phi_\lambda\right)$$ Basic identity: $$\left[CF_\lambda(x)\right]^2 + \left[SF_\lambda(x)\right]^2 = \frac{4}{4+\lambda^2}$$	Hyperbolic λ-sine: $$sF_\lambda(x) = \frac{\Phi_\lambda^x - \Phi_\lambda^{-x}}{\sqrt{4+\lambda^2}} = \frac{2}{\sqrt{4+\lambda^2}}sh\left(x\ln\Phi_\lambda\right)$$ Hyperbolic $\lambda-$cosine: $$cF_\lambda(x) = \frac{\Phi_\lambda^x + \Phi_\lambda^{-x}}{\sqrt{4+\lambda^2}} = \frac{2}{\sqrt{4+\lambda^2}}ch\left(x\ln\Phi_\lambda\right)$$ Basic identity: $$\left[cF_\lambda(x)\right]^2 - \left[sF_\lambda(x)\right]^2 = \frac{4}{4+\lambda^2}$$

Parametric forms of the basic surfaces based on $\lambda-$sines and $\lambda-$cosines

$$X = R\frac{\sqrt{4+\lambda^2}}{2}SF_\lambda(u')\cos(v')$$ $$Y = R\frac{\sqrt{4+\lambda^2}}{2}SF_\lambda(u')\sin(v')$$ $$Z = R\frac{\sqrt{4+\lambda^2}}{2}CF_\lambda(u')$$	$$X = R\frac{\sqrt{4+\lambda^2}}{2}sF_\lambda(u')\cos(v')$$ $$Y = R\frac{\sqrt{4+\lambda^2}}{2}sF_\lambda(u')\sin(v')$$ $$Z = R\frac{\sqrt{4+\lambda^2}}{2}cF_\lambda(u')$$
Spherical metric forms of the Gaussian curvature $K = \frac{1}{R^2} > 0$: $$(ds)^2 = R^2\left[\begin{array}{c}\left(\ln^2\Phi_\lambda(du')\right)^2 + \\ +\frac{4+\lambda^2}{4}\left(SF_\lambda(u')\right)^2(du')^2\end{array}\right]$$ $$\frac{\pi k}{\ln\Phi_\lambda} < u' < \frac{\pi(k+1)}{\ln\Phi_\lambda},$$ $$-\infty < v < +\infty\ (k \in Z),\ \lambda > 0$$	Lobachevski's hyperbolic metric forms of the Gaussian curvature $K = -\frac{1}{R^2} < 0$: $$(ds)^2 = R^2\left[\begin{array}{c}\left(\ln^2\Phi_\lambda(du')\right)^2 + \\ +\frac{4+\lambda^2}{4}\left(sF_\lambda(u')\right)^2(du')^2\end{array}\right]$$ $$0 < u' < +\infty,$$ $$-\infty < v < +\infty\ (k \in Z),\ \lambda > 0$$

Spherical solution	Pseudo - spherical solution
Connection between spherical and hyperbolic λ-functions	
$$SF_\lambda(x) = \frac{\sin(x\ln\Phi_\lambda)}{sh(x\ln\Phi_\lambda)}sF_\lambda(x);$$ $$sF_\lambda(x) = \frac{sh(x\ln\Phi_\lambda)}{\sin(x\ln\Phi_\lambda)}SF_\lambda(x);$$	$$CF_\lambda(x) = \frac{\cos(x\ln\Phi_\lambda)}{ch(x\ln\Phi_\lambda)}cF_\lambda(x)$$ $$cF_\lambda(x) = \frac{ch(x\ln\Phi_\lambda)}{\cos(x\ln\Phi_\lambda)}CF_\lambda(x)$$

Distances between metric forms

Distance between the standard spherical metric form and spherical metric λ-form: $\rho_{12} = R\left|(\ln\Phi_\lambda)-1\right|, \lambda > 0$ **Normalized distance :** $\bar{\rho}_{12} = \frac{\rho_{12}}{R} = \left|(\ln\Phi_\lambda)-1\right|, \lambda > 0$ At $\bar{\rho}_{12} = 0 \Rightarrow \lambda^* = 2sh(1) \Rightarrow$ $\Rightarrow \Phi_{\lambda^*} = e \Rightarrow \ln\Phi_{\lambda^*} = 1,$ the spherical metric form consides with the standard spherical metric form; at $\bar{\rho}_{12} > 0$ these forms do not coinside	Distance between the Lobachevski's hyperbolic metric form and Lobachevski's metric λ-form: $\rho_{12} = R\left|(\ln\Phi_\lambda)-1\right|, \lambda > 0$ **Normalized distance :** $\bar{\rho}_{12} = \frac{\rho_{12}}{R} = \left|(\ln\Phi_\lambda)-1\right|, \lambda > 0$ At $\bar{\rho}_{12} = 0 \Rightarrow \lambda^* = 2sh(1) \Rightarrow$ $\Rightarrow \Phi_{\lambda^*} = e \Rightarrow \ln\Phi_{\lambda^*} = 1,$ Lobachevski's metric form consides with Lobachevski's hyperbolic metric λ-form; at $\bar{\rho}_{12} > 0$ these forms do not coinside

The minimal distance

Comparing the *metallic proportions* by the *normalized distance*, we lead to the conclusion, that the *silver proportion* $1+\sqrt{2} \approx 2.414$ $(\lambda = 2)$ has the **minimal distance** $\bar{\rho}_{12} = 0.118$ among all other metallic proportions Φ_λ for the positive integer values of the parameter $\lambda = 1, 2, 3, ...$

This means that among the *spherical metric* λ-forms, generated by the metallic proportions Φ_λ for the integer values of $\lambda = 1, 2, 3, ...$, the "*silver*" *spherical metric* λ-*form*, generated by the *silver proportion* $1+\sqrt{2} \approx 2.414$ $(\lambda = 2)$, is the closest spherical metricl λ-form to the standard metric spherical form, in the sense of the normalized distance $\bar{\rho}_{12}$.

Unique properties of the metric λ – forms

Diffeomorphism $f : u = (\ln \Phi_\lambda)u', v = v', \lambda > 0$ is a mapping, under the action of which

the standard spherical metric form (Gaussian curvature $K = \dfrac{1}{R^2} > 0, \bar{\rho}_{12} = 0$)

is transformed into the spherical λ-form (Gaussian curvature $K = \dfrac{1}{R^2} > 0, \bar{\rho}_{12} > 0$)

and Lobachevski's metric form (Gaussian curvature $K = -\dfrac{1}{R^2} < 0, \bar{\rho}_{12} = 0$) is

transformed into Lobachevskyi's hyperbolic metric λ-forms

(Gaussian curvature $K = -\dfrac{1}{R^2} < 0, \bar{\rho}_{12} > 0$).

Under the action of the mapping f when going from $\bar{\rho}_{12} = 0$ to $\bar{\rho}_{12} > 0$, the metric properties of **isometric equivalence** and **equireality** of the metric forms are **saved**. Under the action of the mapping f when going from $\bar{\rho}_{12} = 0$ to $\bar{\rho}_{12} > 0$ the individual to the metric properties of **isometric identity** and **conformity** are **not saved**.

Step 5

The main results of the study

The spherical solution of Hilbert's Fourth Problem	The hypebolic solution of Hilbert's Fourth Problem
1.There are infinite geometries, induced by the spherical metric λ- forms of the same constant positive curvature.	1. There are infinite geometries, induced by the hyperbolic metric λ- forms of the same constant negative curvature.
2. Among the *spherical metric λ – forms*, generated by the metallic proportions Φ_λ for the integer values of $\lambda = 1, 2, 3, ...,$ the *"silver" spherical metric λ – form*, generated by the *silver proportion* $1 + \sqrt{2} \approx 2.414$ $(\lambda = 2)$, is the closest spherical metric λ – form to the standard metric spherical form, in the sense of the normalized distance $\bar{\rho}_{12}$.	2. Among the *hyperbolic metric λ – forms*, generated by the metallic proportions Φ_λ for the integer values of $\lambda = 1, 2, 3, ...,$ the *"silver" hyperbolic metric λ – form*, generated by the *silver proportion* $1 + \sqrt{2} \approx 2.414$ $(\lambda = 2)$, is the closest spherical metric λ – form to the standard metric spherical form, in the sense of the normalized distance $\bar{\rho}_{12}$.
3. The silver sperical geometry is of great interest for theoretical physics and can realy exist in physical world.	3. The silver hyperbolic geometry is of great interest for theoretical physics and can realy exist in physical world.

6.5. The general conclusions for the book

1. Discussing the history of mathematics and the development of new mathematical ideas and theories, we should draw a particular attention to the great role of Euclid's *Elements* in this process. Academician **Andrey Kolmogorov** identifies several stages in the development of mathematics [2]. According to Kolmogorov, the modern period in mathematics was starting in 19th century. **Expanding the object of mathematics became the most significant feature of the 19th century mathematics.** At the same time, according to Kolmogorov, the creation of Lobachevski's "imaginary geometry" became *"a remarkable example of the theory that has arisen as a result of internal development of mathematics ... It is an example of the geometry, which overcame a belief in the permanence of the axioms, consecrated millennial development of mathematics, and gave a comprehension of the possibility of creating significant new mathematical theories ...* [2]. As we know, the "hyperbolic geometry" in its origins dates back to **Euclid's** 5th postulate. For several centuries, from **Ptolemy** and **Proclus**, mathematicians tried to prove this postulate. First a brilliant solution to the problem was given by Russian mathematician **Nikolay Lobachevski** in the first half of the 19th century. This became the beginning of contemporary stage in the development of mathematics.

2. At the turn of the 19th and 20th century, the eminent mathematician **David Hilbert** formulated 23 mathematical problems, which largely stimulated the development of mathematics in the 20 century. One of them (Hilbert's Fourth Problem) refers directly to the hyperbolic geometry. Hilbert put forward for mathematicians the following fundamental problem [6]: *"The more general question now arises: whether from other suggestive standpoints geometries may not be devised which, with equal right, stand next to Euclidean geometry."* Hilbert's quote contains the formulation of a very important scientific problem, which has a fundamental importance not only for mathematics, but also for all theoretical natural sciences: are there non-Euclidean geometries, which are close to the Euclidean geometry and are interesting from the "other suggestive standpoints?" If we consider it in the context of theoretical natural sciences, then Hilbert's Fourth Problem is about searching **NEW HYPERBOLIC WORLDS OF NATURE**, which are close to the Euclidean geometry and reflect some new properties of

Nature's structures and phenomena. Unfortunately, the efforts of mathematicians to solve this problem did not lead to substantial progress. In modern mathematics there is no consensus about the solution of this problem. In mathematical literature Hilbert's Fourth Problem is considered as formulated **very vague** what makes difficult its final solution. As it is noted in Wikipedia [8], *"the original statement of Hilbert, however, has also been judged too vague to admit a definitive answer."*

3. Besides of 5th postulate, Euclid's Elements contain another fundamental idea that permeates the entire history of science. We are talking about the "idea of the Universe Harmony," which in ancient Greece was associated with the "golden ratio" and Platonic solids. "Proclus hypothesis," formulated in the 5th century AD by the Greek philosopher and mathematician **Proclus Diadochus** (412 – 485), contains the unexpected view on Euclid's *Elements*. According to Proclus, Euclid's main goal was to build the full theory of regular polyhedra ("Platonic solids"). This theory was outlined by Euclid in the Book XIII, that is, in the concluding book of the *Elements* what in itself is an indirect confirmation of "Proclus' hypothesis." To solve this problem, Euclid had included the necessary mathematical information into the *Elements*. **This mathematical information was used by Euclid to solve the main problem - the creation of a complete theory of the Platonic solids.** The most curious thing is that he had introduced in the Book II the "golden ratio" for the creation of geometric theory of the dodecahedron.

4. Starting from Euclid, the "golden section" and "Platonic solids" permeate with the "red thread" the history of mathematics and theoretical natural sciences. In modern science, Platonic solids have become a source for new scientific discoveries, particularly of **fullerenes** (Nobel Prize in chemistry, 1996) and **quasi-crystals** (Nobel Prize in chemistry, 2011). The publication of **Stakhov's book "The Mathematics of Harmony. From Euclid to Contemporary Mathematics and Computer Science"** [19] is a reflection of the very important trend in the development of modern science (including mathematics), the revival of "harmonic ideas" of Pythagoras, Plato and Euclid.

5. The "metallic proportions" by **Vera W. de Spinadel** [52], which are a generalization of the classical "golden section," are a new class of mathematical constants, representing fundamental theoretical and practical

significance. Besides of Argentinean mathematician Vera W. de Spinadel, many researchers from different countries and continents (French mathematician **Midhat Gazale** [53], American mathematician **Jay Kappraff** [54], Russian engineer **Alexander Tatarenko** [55], Armenian philosopher and physicist **Hrant Arakelyan** [56], Russian researcher **Victor Shenyagin** [57], Ukrainian physicist **Nikolai Kosinov** [58], Spanish mathematicians **Falcon Sergio** and **Plaza Angel** [59] and others) independently came to one and the same mathematical proportions called the "metallic proportions" [52]. All this confirms the fact that the appearance of new (harmonic) mathematical constants matured in mathematics.

6. The new classes of hyperbolic functions, based on the "golden section" and Fibonacci numbers (hyperbolic Fibonacci and Lucas functions), became one of the most important results of the "mathematics of harmony" [19], having direct relevance to the hyperbolic geometry. First this mathematical result was obtained by the Ukrainian mathematicians **Alexey Stakhov, Ivan Tkachenko, Boris Rosin** [9-12]. The hyperbolic Fibonacci and Lucas λ-functions, based on Spinadel's "metallic proportions," became very important step in the creation of general theory of the "harmonic" hyperbolic functions [13, 14].

7. Researches of the Ukrainian architect Oleg Bodnar are a significant step in the development of "hyperbolic geometry". Bodnar showed [20] that a special kind of the hyperbolic geometry, based on the "golden" hyperbolic functions, has wide distribution in wild Nature and underlies the botanic phenomenon of phyllotaxis (pine cones, cacti, pineapples, sunflower heads, etc.). **The discovery of the Ukrainian architect Oleg Bodnar showed that hyperbolic geometry is much more common in the wildlife than it seemed before. Perhaps, the entire wildlife is the epitome of "Bodnar's geometry."**

8. From this point of view, the original solution of Hilbert's Fourth Problem, based on Stakhov's "Mathematics of Harmony" [19], in particular, on the "metallic proportions" [52] and hyperbolic Fibonacci λ-functions [13,14], is of special interest for mathematics and all theoretical natural sciences. This is the main result of this book and the previous works of the authors in this area [29-32]. It is proved in these works that there is an infinite number of new hyperbolic geometries, which *"with equal right, stand*

next to Euclidean geometry" (**David Hilbert**). This solution of Hilbert's Problem puts forward in front to theoretical natural sciences (physics, chemistry, biology, genetics and so on) the scientific problem to search new ("harmonic") worlds of Nature, which can objectively exist in the world around us. In this regard, we should draw a special attention to the fact that the new hyperbolic geometry, based on the "silver" hyperbolic functions with the base $\Phi_2 = 1 + \sqrt{2} \approx 2.41$, is the closest to Lobachevski's geometry, based of the classical hyperbolic functions with the base $e \approx 2.71$. Its distance to the Lobachevski's geometry is equal $\overline{\rho}_{12} \approx 0.1677$ what is the smallest among all the distances for Lobachevski's metric (λ, μ)-forms.

We may assume that the **"silver" hyperbolic functions** and the generated by them **"silver" hyperbolic geometry** can be soon be found in nature after "Bodnar's geometry," based on the **"golden" hyperbolic functions** with the base $\Phi = \frac{1 + \sqrt{5}}{2} \approx 1.618$.

9. . The second result of the book is the following. In this book, the authors have attempted to spread the approach, used in [29-32], on the *spherical geometry*. To solve this problem, a new class of elementary functions, called **spherical Fibonacci λ-functions**, was introduced in this book. This class of functions allowed to solve Hilbert's Fourth Problem with respect to the *spherical* geometry.

10. A study in this book is of considerable interest from the standpoint of the mathematics history and prospects of its development in close association with theoretical natural sciences. This study unites the ancient "golden section," described in Euclid's Elements, with Lobachevski's hyperbolic geometry and spherical geometry. This unexpected union led to the original solution of Hilbert's Fourth Problem for hyperbolic and spherical geometry what is the very important mathematical result, which is starting the new ("harmonic") stage in the development of non-Euclidean geometry and its applications in theoretical natural sciences.

11. **Taking into consideration "Dirac's Principle of Mathematical Beauty" and looking from this point of view on the spherical and hyperbolic Fibonacci functions, as well as "Bodnar geometry," which got the widest distribution in wildlife (botanical phenomenon of phyllotaxis), the authors believe that, if David Hilbert would live in the moment, then he would have given his preference to solving the Fourth Problem**

in terms of the "mathematics of harmony," because, in this "harmonic" solution, Leibniz's concept about "pre-established harmony" united with "Dirac's Principle of Mathematical Beauty" as the initial principle of physical theory. The new solution to Hilbert's Fourth Problem is opening vast opportunities for all theoretical natural sciences and puts forward the challenge finding new hyperbolic and spherical worlds of Nature.

References

1. "Hyperbolic geometry." From Wikipedia, the free encyclopaedia
   http://en.wikipedia.org/wiki/Hyperbolic_geometry

2. A.N. Kolmogorov. "Mathematics in Its Historical Development (in Russian)." Moscow: Nauka, 1991.

3. Non-Euclidean geometry. From Wikipedia, the free encyclopaedia (Russian)
   http://ru.wikipedia.org/wiki/%D0%9D%D0%D0%B5%D0%B5%D0%B2%D0%B
   A%D0%BB%D0%B8%D0%B4%D0%BE%D0%B2%D0%B0_%D0%B3%D
   0%B5%D0%BE%D0%BC%D0%B5%D1%82%D1%80%D0%B8%D1%8F

4. Spherical geometry. From Wikipedia, the free encyclopaedia
   http://en.wikipedia.org/wiki/Spherical_geometry

5. B.A. Dubrovin, S.P. Novikov, A.T. Fomenko, "Modern Geometry. Methods and Applications," Moscow: Nauka , 1979 (in Russian).

6. Lecture "Mathematical Problems" by Professor David Hilbert
   http://aleph0.clarku.edu/~djoyce/hilbert/problems.html#prob4

7. "Hilbert's Problems" (under P.S. Alexandrov's edition) (in Russian), Moscow: Nauka, 1969.

8. "Hilbert's Fourth Problem." From Wikipedia. The Free Encyclopaedia.
   http://en.wikipedia.org/wiki/Hilbert's_fourth_problem

9. A.P. Stakhov, I.S. Tkachenko. "Hyperbolic Fibonacci Trigonometry (in Russian)." *Reports of the Ukrainian Academy of Sciences*, Vol. 208, No 7, 1993, 9-14.

10. A. Stakhov, B. Rozin. "On a new class of hyperbolic function." *Chaos, Solitons & Fractals*, 2004, 23, 379-389.

11. A. Stakhov, B. Rozin. "The "golden" hyperbolic models of Universe." *Chaos, Solitons & Fractals*, 2007, Vol. 34, Issue 2, 159-171.

12. A.P. Stakhov, B.N. Rozin. "The Golden Section, Fibonacci series and new hyperbolic models of nature." *Visual Mathematics*, Vol. 8, No 3, 2006
    http://www.mi.sanu.ac.rs/vismath/stakhov/index.html

13. A.P. Stakhov. "Gazale formulas, a new class of the hyperbolic Fibonacci and Lucas functions, and the improved method of the «golden» cryptography." *Academy of Trinitarizam*. Moscow: Electronic number 77-6567, publication 14098, 21.12.2006, http://www.trinitas.ru/rus/doc/0232/004a/02321063.htm

14. Alexey Stakhov. "On the general theory of hyperbolic functions based on the hyperbolic Fibonacci and Lucas functions and on Hilbert's Fourth Problem," *Visual Mathematics*, Vol.15, No.1, 2013 http://www.mi.sanu.ac.rs/vismath/pap.htm)

15. Charles H. Kann. Pythagoras and Pythagoreans. A Brief History. Hackett Publishing Co, Inc., 2001.

16. Leonid Zhmud. The origin of the History of Science in Classical Antiquity. Published by Walter de Gruyter, 2006.

17. Craig Smorinsky. History of Mathematics. A Supplement. Springer, 2008.

18. A.P. Stakhov. "The Mathematics of Harmony: Clarifying the Origins and Development of Mathematics." *Congressus Numerantium*, VOLUME CXCIII, December, 2008, 5-48

19. A.P. Stakhov. "The Mathematics of Harmony. From Euclid to Contemporary Mathematics and Computer Science". International Publisher "World Scientific" (New Jersey, London, Singapore, Beijing, Shanghai, Hong Kong, Taipei, Chennai), 2009.

20. O. Y. Bodnar. "The Golden Section and Non-Euclidean Geometry in Nature and Art (in Russian). Lvov: Publishing House "Svit", 1994.

21. O. Bodnar. "Dynamic Symmetry in Nature and Architecture." *Visual Mathematics*, 2010, Vol.12, No.4. http://www.mi.sanu.ac.rs/vismath/BOD2010/index.html

22. O. Bodnar. "Geometric Interpretation and Generalization of the Non-classical Hyperbolic Functions." *Visual Mathematics*, 2011, V.13, No.2 http://www.mi.sanu.ac.rs/vismath/bodnarsept2011/SilverF.pdf

23. O. Bodnar. "Minkovski's Geometry in the Mathematical Modeling of Natural Phenomena." *Visual Mathematics*, 2012, Vol.14, No.1. http://www.mi.sanu.ac.rs/vismath/bodnardecembar2011/mink.pdf

24. "Fullerene." From Wikipedia, the free encyclopaedia http://en.wikipedia.org/wiki/Fullerene

25. "Quasi-crystal." From Wikipedia, the free encyclopaedia
http://en.wikipedia.org/wiki/Quasicrystal

26. E.M. Soroko. "Structural Harmony of Systems (in Russian)." Minsk: Publishing House "Nauka i Tekhnika," 1984.

27. Hilbert's Tenth Problem. From Wikipedia, the free Encyclopedia
http://en.wikipedia.org/wiki/Hilbert's_tenth_problem

28. S.V. Petoukhov "Metaphysical aspects of the matrix analysis of genetic code and the golden section." *Metaphysics: Century XXI (in Russian).* Moscow: Publishing House "BINOM", 2006, 216-250.

29. A. Stakhov, S. Aranson. "Hyperbolic Fibonacci and Lucas Functions, "Golden" Fibonacci Goniometry, Bodnar's Geometry, and Hilbert's Fourth Problem. Part I." *Applied Mathematics*, Vol.2, No.1, January 2011, 74-84.
http://www.scirp.org/journal/am/

30. A. Stakhov, S. Aranson. "Hyperbolic Fibonacci and Lucas Functions, "Golden" Fibonacci Goniometry, Bodnar's Geometry, and Hilbert's Fourth Problem. Part II." *Applied Mathematics*, Vol.2, No.2, February 2011, 181-188. http://www.scirp.org/journal/am/

31. A. Stakhov, S. Aranson. "Hyperbolic Fibonacci and Lucas Functions, "Golden" Fibonacci Goniometry, Bodnar's Geometry, and Hilbert's Fourth Problem. Part III.. *Applied Mathematics*, Vol.2, No. 3, March 2011, 283–293.
http://www.scirp.org/journal/am/

32. A.P. Stakhov. "Hilbert's Fourth Problem: Searching for Harmonic Hyperbolic Worlds of Nature." *Applied Mathematics and Physics*, Vol.1, No.3, 2013, 60-66 http://www.scirp.org/journal/jamp/.

33. Academician Mitropolsky's commentary on the scientific research of the Ukrainian scientist Doctor of Engineering Sciences Professor Alexey Stakhov. In the book byAlexey Stakhov "The Mathematics of Harmony. From Euclid to Contemporary Mathematics and Computer Science". (World Scientific), 2009.

34. Harmony of spheres. The Oxford dictionary of philosophy. Oxford University Press, 1994, 1996, 2005.

35. Musica universalis. From Wikipedia, the free encyclopaedia
http://en.wikipedia.org/wiki/Musica_universalis

36. Vladimir Dimitrov. A new kind of social science. Study of self-organization of human dynamics. Morrisville Lulu Press, 2005.

37. A.P. Stakhov. The Golden Section and Modern Harmony Mathematics. Applications of Fibonacci Numbers, Kluwer Academic Publishers, Vol. 7, 1998. 393 – 399.

38. N.N. Vorobyov. Fibonacci Numbers. Moscow: Publishing House "Nauka", 1961 (in Russian).

39. V.E. Hoggat. Fibonacci and Lucas Numbers. - Palo Alto, CA: Houghton-Mifflin; 1969.

40. S. Vajda. Fibonacci & Lucas Numbers, and the Golden Section. Theory and Applications. Ellis Horwood limited; 1989.

41. F. Klein, Lectures on the Icosahedron. Courier Dover Publications, 1956.

42. A.P. Stakhov. The generalized golden proportions and a new approach to geometric definition of a number. Ukrainian Mathematical Journal 2004, 56: 1143-1150.

43. G.D. Grimm. Proportionality in Architecture. Leningrad-Moscow: Publishing House "ONTI", 1935 (in Russian).

44. A.J. Khinchin. Continued Fractions. Moscow: Fizmathiz, 1960.

45. Jan Grzedzielski. Energetycno-geometryczny kod Przyrody. Warszawa, Warszwskie centrum studenckiego ruchu naukowego, 1986 (in Polen).

46. D. Gratia. Quasi-crystals. Uspekhi physicheskikh nauk, 1988, Vol. 156, No. 2 (in Russian).

47. G. Bergman. A number system with an irrational base. Mathematics Magazine, 1957, No.31: 98-119.

48. H.W. Gould. A history of the Fibonacci Q-matrix and higher-dimensional problems. The Fibonacci Quarterly, 19, 1981, 250-257.

49. V.G. Shervatov. Hyperbolic Functions. Moscow: Fizmatgiz, 1958 (in Russian).

50. A.P. Stakhov. Codes of the Golden Proportion. Moscow: Publishing House "Radio and Communication," 1984 (in Russian).

51. V. Arnold. A complexity of the finite sequences of zeros and units and geometry of the finite functional spaces. Lecture at the session of the Moscow Mathematical Society, May 13, 2007. http://elementy.ru/lib/430178/430281

52. Vera W. de Spinadel. "From the Golden Mean to Chaos." Nueva Libreria, 1998 (second edition, Nobuko, 2004).

53. Midhat J. Gazale. "Gnomon. From Pharaohs to Fractals." Princeton, New Jersey: Princeton University Press, 1999.

54. Jay Kappraff. "Connections. The geometric bridge between Art and Science." Second Edition. Singapore, New Jersey, London, Hong Kong: World Scientific, 2001.

55. Alexander Tatarenko. "The golden T_m-harmonies' and D_m-fractals (in Russian). *Academy of Trinitarism*. Moscow: № 77-6567, Electronic publication 12691, 09.12.2005 http://www.trinitas.ru/rus/doc/0232/009a/02320010.htm

56. Hrant Arakelyan. "The numbers and magnitudes in modern physics (in Russian)." Yerevan, Publishing House "Armenian Academy of Sciences," 1989.

57. V.P. Shenyagin. "Pythagoras, or how everyone creates his own myth. The fourteen years after the first publication of the quadratic mantissa's proportions.(in Russian)." *Academy of Trinitarism*. Moscow: № 77-6567, electronic publication 17031, 27.11.2011. http://www.trinitas.ru/rus/doc/0232/013a/02322050.htm

58. N.V. Kosinov. "The golden ratio, golden constants, and golden theorems (in Russian). *Academy of Trinitarism*. Moscow: № 77-6567, electronic publication 14379, 02.05.2007. http://www.trinitas.ru/rus/doc/0232/009a/02321049.htm

59. Sergio Falcon, Angel Plaza. "On the Fibonacci k-numbers." *Chaos, Solitons & Fractals*, Vol. 32, Issue 5, June 2007 : 1615-1624.

60. A.P Stakhov. "A generalization of the Cassini formula." Visual Mathematics, Vol. 14, No.2, 2012 http://www.mi.sanu.ac.rs/vismath/stakhovsept2012/cassini.pdf

61. A.A. Andronov, A.A. Vitt, S.E. Khaikin. "Theory of oscillations," Moscow: Fizmatgiz, 1959 (in Russian).

62. N.N. Bautin, E.A. Leontovich. "Methods and ways of qualitative study of dynamic systems on a plane," Moscow: Nauka, 1976 (in Russian).

63. Ju. I. Neimark, "Methods of point mappings in theory of non-linear oscillations," Moscow: Nauka, 1972 (in Russian).

64. S.Kh. Aranson. "Qualitative Properties of Foliations on Closed Surfaces," *Journal of Dynamical and Control Systems*, New-York and London: Plenum Press, Vol.6, No 1, 2000, 127-157.

65. S.Kh. Aranson, E.V. Zhuzoma. "Nonlocal Properties of Analytic Flows on Closed Orientable Surfaces," *Proceedings of the Steklov Institute of Mathematics*. Vol. 244, 2004, 2-17.

66. S.Aranson, V.Medvedev, E.Zhuzhoma "Collapse and Continuity of Geodesic Frameworks of Surface Foliations." In the book *"Methods of Qualitative Theory of Differential Equations and Related Topics"* (Dedicated to the memory of E.A. Andronov-Leontovich), USA: American Mathematical Society. 35-49

67. S.Kh. Aranson, Belitsky, E.V. Zhuzhoma, "Introduction to the QualitativeTheory of Dynamical Systems on surfaces," *USA : American Mathematical Society*, Vol.153, 1996.

68. D.V. Anosov, S.Kh. Aranson, V.I. Arnold, I.U. Bronshtein, V.Z. Grines, Yu.S. Il'yashenko, "Ordinary Differential Equations and Smooth Dynamical Systems," Berlin: Springer, 1997.

69. D.V. Anosov, Aranson S.Kh, V.Z.Grines, R.V.Plykin, A.V.Safonov , E.A.Sataev, S.V.Shlyachkov,V.V.Solodov,A.N.Starkov,A.M.Stepin, "Dynamical Systems with Hyperbolic Behaviour," *Encyclopaedia of Mathematical Sciences. Dynamical Systems IX*, Vol.66 , Berlin: Springer, 1995, 1-235.

70. H. Busemann, "On Hilbert's Fourth Problem (in Russian)." *Uspechi mathematicheskich nauk,* 1966, Vol.21, No 1(27), 155-164.

71. A.V. Pogorelov. "Hilbert's Fourth Problem (in Russian)." Moscow: Nauka, 1974.

72. Hilbert's Problems. From Wikipedia, the free Encyclopaedia http://en.wikipedia.org/wiki/Hilbert's_problems

73. S.Kh. Aranson. "Once again on Hilbert's Fourth Problem (in Russian). *Academy of Trinitarizm*. Moscow, № 77-6567, publication 15677,01.12.2009. http://www.trinitas.ru/rus/doc/0232/009a/02321180.htm

74. V.I. Arnold, Yu.S. Il'yashenko, D.V.Anosov, I.U. Bronshtein, S.Kh.Aranson, V.Z.Grines, «Dynamical Systems I», *Encyclopaedia of Mathematical Sciences,* Vol.1 , Berlin: Springer- Verlag, 1998.

75. Yandell Benjamin H. "The Honors class- Hilbert's problems and Their Solvers," Massachusetts: A.K.Peters Naticn, 2003.

76. AP. Stakhov. "Non-Euclidean Geometries. From the "game of postulates" to the "game of function (in Russian)." *Academy of Trinitarizm*. Moscow, № 77-6567, publication 18048, 9.05.2013 (Russian). http://www.trinitas.ru/rus/doc/0016/001d/00162125.htm

About the authors

Alexey Stakhov is Doctor of Science in Computer Science, Professor. Author of over 500 scientific publications, including 10 books and more than 65 foreign patents (U.S., Japan, England, France, Germany, Canada and other countries). He is one of the world leaders in the field of the "golden section" and "mathematics of harmony." The book "The Mathematics of Harmony. From Euclid to Contemporary Mathematics and Computer Science "(World Scientific, 2009) is his main scientific achievement. The book is a reflection of the most important trends in the development of modern science, a return to the "harmonic ideas" of Pythagoras, Plato and Euclid. Since 2003 he is President of the International Club of the Golden Section, and since 2005 he is Director of the Institute of the Golden Section, Academy of Trinitarism (Russia). Professor Stakhov was Visiting Professor of many famous universities (Vienna University of Technology, Austria, 1976; Fridrich Shiller's University of Jena, DDR, 1986, Dresden University of Technology, Dresden, DDR, 1988).

Samuil Aranson, Doctor of Physical and Mathematical Sciences (differential equations, geometry and topology), Professor. Professor Aranson is widely known in modern mathematics. In 1995 according to the Decree of the President of Russia he awarded the honorary title "Honoured Worker of Science of Russia." In 1997 he was elected an academician of the Russian Academy of Natural Sciences. Scientific activities of Professor Aranson closely linked with Gorky (Nizhny Novgorod) school of nonlinear oscillations, organized by the academician A.A. Andronov and refers to the qualitative theory of ordinary differential equations. Professor Aranson is the author of many original books and articles in mathematics, published during from 1963 to 2006. He published (individually or collectively with other authors) more than 200 scientific works, including monographs, published in Russia, USA, Germany and other countries.. He worked as Visiting professor in the leading Israel universities (University of Jerusalem, Tel Aviv University and other universities).

Made in the USA
Lexington, KY
18 July 2014